Kitchen & Bathroom Plumbing

Ron Hazelton, chief consultant for **HOW TO FIX IT**, is the Home Improvement Editor for ABC-TV's *Good Morning America* and host of his own home improvement series, Ron Hazelton's *House-Calls.* He has produced and hosted more than 200 episodes of *The House Doctor,* a home-improvement series airing on the *Home and Garden Television Network* (HGTV).

On television, and in real life, Ron is a coach who visits people in their own homes, helping them do things for themselves. He pioneered the concept of on-location, home-improvement television, making over 600 televised house calls, doing real-life projects with real people.

The son of a building contractor, Ron has always had a fascination with the home and how it works. He left a successful career as a marketing executive to learn woodworking, eventually becoming a Master Craftsman and cabinetmaker. In 1978, he founded Cow Hollow Woodworks in San Francisco, an antique restoration workshop that restored over 17,000 pieces of furniture during his tenure.

Matthew Ruggiero, a Master Tradesman, is the founder of Ruggiero Plumbing & Heating, Inc., a family owned and operated plumbing and heating company located in New York. In business for 40 years, the Ruggieros are plumbing and heating contractors, specializing for the past 15 years on servicing municipal utility lines.

Evan Powell is the Director of Engineering and Services at Southeastern Products, Inc. The author of *The Complete Guide to Home Appliance Repair*-and many other widely acclaimed books on home repair, Mr. Powell is also a regular contributor to numerous industry publications. He appears frequently on television as an expert on home repair and consumer topics.

Other Publications:

Do It Yourself
Home Repair and Improvement
The Time-Life Complete Gardener
The Art of Woodworking

Cooking
Weight Watchers™ Smart Choice
   Recipe Collection
Great Taste/Low Fat
Williams-Sonoma Kitchen Library

History
The American Story
Voices of the Civil War
The American Indians
Lost Civilizations
Mysteries of the Unknown
Time Frame
The Civil War
Cultural Atlas

Time-Life Kids
Library of First Questions and Answers
A Child's First Library of Learning
I Love Math
Nature Company Discoveries
Understanding Science & Nature

Science/Nature
Voyage Through the Universe

For information on and a full description of any of the Time-Life Books series listed above, please call 1-800-621-7026 or write:

Reader Information
Time-Life Customer Service
P.O. Box C-32068
Richmond, Virginia 23261-2068

HOW TO FIX IT

Kitchen & Bathroom Plumbing

By The Editors of Time-Life Books, Alexandria, Virginia

With **TRADE SECRETS** From **Ron Hazelton**

Contents

FIX IT: Faucets

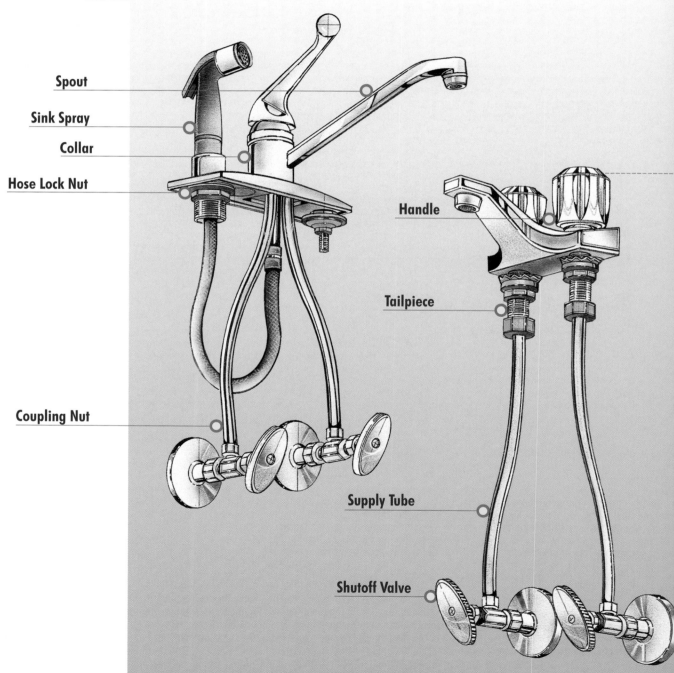

Spout

Sink Spray

Collar

Hose Lock Nut

Handle

Tailpiece

Coupling Nut

Supply Tube

Shutoff Valve

Chapter 1

Contents

How They Work

Sink faucets contain simple valves that control the flow of thousands of gallons of water each year in the kitchen and bathroom. Two major types, either of which may be found in both rooms, are shown at left. One of them has two handles set about four inches apart; a variation has wider spacing between the handles. Single-lever faucets comprise the other type. The one shown here, intended for the kitchen, is equipped with a spray hose.

Either type of faucet may have one of several kinds of valves, whose repair is explained beginning on page 10. Regardless of type, a faucet is fed from below by hot- and cold-water supply tubes, each controlled by a shutoff valve that makes it possible to work on the faucet without interrupting the flow of water to other fixtures in the house.

Troubleshooting

Problem	Solution
• **Faucet drips from spout**	Compression faucet: Replace washer if worn or damaged **11** • Replace pitted or corroded seat **11** •
• **Faucet leaks from handle**	Compression faucet: Replace worn or damaged O-rings (newer models) **11** • Tighten the packing nut, replace worn packing, or replace a bent stem (older models) **13** • Reverse compression faucet: Replace worn packing washer **15** • Diaphragm faucet: Replace worn O-ring **17** • Disk faucet: Replace worn O-ring on disk assembly **19** •
• **Faucet handle loose**	All double-handle faucets: Tighten handle screw, if any
• **Faucet leaks from collar**	Rotating-ball faucet: Replace worn O-rings **22** • Cartridge faucet: Replace worn O-rings **25** •
• **Faucet leaks around base or has reduced flow**	Disk faucet: Replace cracked or pitted disk assembly **19** • Replace worn inlet seals **19** •
• **Flow from spout reduced**	Clean the aerator **33** •
• **Water under sink**	Tighten faucet-set lock nuts under the sink **31** • Replace putty or gasket **29** • Replace worn faucet **29** • Replace leaky supply tubes **28** •
• **Aerator leaks around edge**	Replace washer in aerator **33** •
• **Spray hose leaks or has reduced flow at spray head**	Replace worn O-ring on diverter valve **33** • Replace worn washer at base of spray head **33** • Clean diverter valve and spray head **33** •

Before You Start

To repair a faucet, you often have to know what kind of valve is inside. Sometimes faucets have characteristics that can help you in this discovery.

Begin by looking for a brand name. If you find one, a call to a plumbing supply store sometimes leads to an identification of type. Knowing approximately when the faucet was installed helps, too.

Or you can experiment. Try turning the handle. A handle that rises and falls as you open and close the faucet is probably a type of compression faucet. Compression, reverse-compression, and diaphragm faucets always have double-handles. Among single-lever faucets, one with a handle that rises and falls as you open and close the faucet is one that most likely has a cartridge.

If none of these tips is conclusive, take off the handle and compare what you see to the exploded views that appear on the following pages.

Before You StartTips:

⋯⫶ Before disassembling faucets, turn off the water supply, open the faucets slightly to let water escape and to relieve pressure, then close the sink drain to keep from losing small parts.

⋯⫶ When using a wrench or pliers on chromed or brass surfaces, "soften" the jaws of the tool with a few turns of masking tape to reduce scratching.

TOOLS

Adjustable wrench
Long-nose pliers
Pliers
Screwdriver
Faucet handle puller
Hex wrench
Channel-joint pliers
Basin wrench
Putty knife
Tubing bender

MATERIALS

Tape
Penetrating oil
Plumber's putty
O-rings
Washers
Faucet repair kit
Replacement faucet set
Flexible supply tubes

Compression Faucets (Newer models)

ANATOMY

Hidden beneath a faucet handle is a series of parts that work together to control water flow. The sleeve (not found on all models) separates the stem assembly from the handle. As the spindle moves up within the stem, it lifts the washer from its seat, allowing water to flow to the spout. As the stem turns in the opposite direction, the washer is pressed tightly against the seat to prevent leakage through the spout.

If the spout leaks, suspect a worn or damaged washer, or perhaps a pitted seat. Leaks from the handle, however, indicate problems with the O-ring on newer models *(left)* or with the packing on older models *(page 12)*.

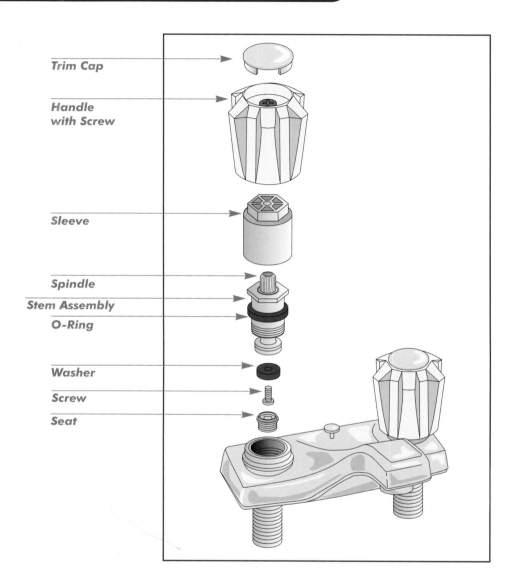

Trim Cap

Handle with Screw

Sleeve

Spindle

Stem Assembly

O-Ring

Washer

Screw

Seat

1. OPENING UP THE FAUCET

• Turn off the water supply, then turn the faucet on, then off to reduce water pressure in the faucet.

• Carefully pry off the trim cap with a knife or small screwdriver *(right)*.

• Remove the handle screw by turning it counterclockwise *(inset)*.

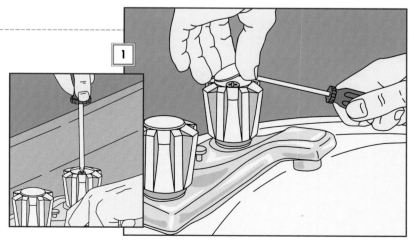

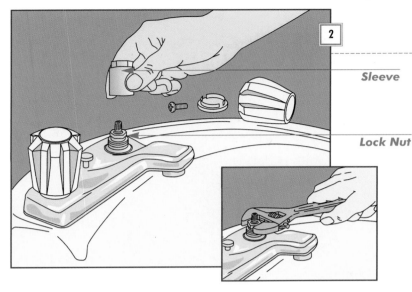

Sleeve

Lock Nut

2. GETTING AT THE STEM

- Lift and remove the faucet handle and sleeve *(left)*. If it is stuck, apply penetrating oil, wait an hour, then tap it gently. As a last resort, use a faucet handle remover (available from plumbing supply stores).

- Remove the lock nut holding the stem to the faucet body *(inset)*.

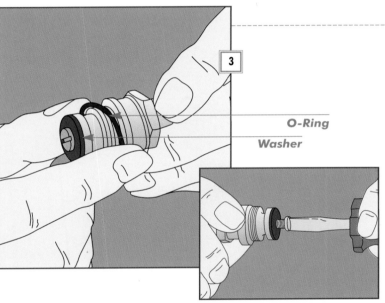

O-Ring

Washer

3. REPLACING THE O-RING AND WASHER

- Grasp the spindle with pliers and lift it out of the faucet body.

- To stop leaks around the handle, remove and replace the O-ring *(left)*.

- To stop leaks around the spout, remove the retaining screw and the washer *(inset)*. If the screw is tight, reinstall the faucet handle for better leverage. Install a new washer, with the flat side against the stem, and secure it with the screw.

- Screw the stem into the faucet body and tighten it snugly with a wrench, then reinstall the handle and test the faucet. If it still leaks, go to Step 4.

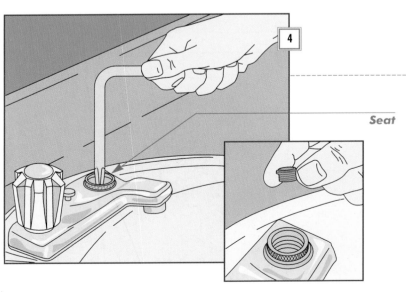

Seat

4. REMOVING THE SEAT

- Remove the stem as before, then unscrew the seat *(left)* with a hex wrench or valve-seat wrench. If the seat is stuck, apply penetrating oil and wait an hour or more before trying again.

- Lift out the old seat *(inset)*.

- Fit an identical replacement seat into the faucet body by hand or with a pair of long-nose pliers and tighten.

Compression Faucets (Older models)

ANATOMY

A four-bladed handle is typical of older style compression faucets. The stem-and-seat design works much as it does in newer compression faucets *(page 10)*, but the faucet relies on packing instead of an O-ring to keep water from leaking around the handle. Packing may be either a rubber or cork washer, graphite-impregnated string, or even plain string. Before disassembling the faucet to replace the packing, however, try tightening the packing nut— sometimes that's all it takes to stop a leaking handle.

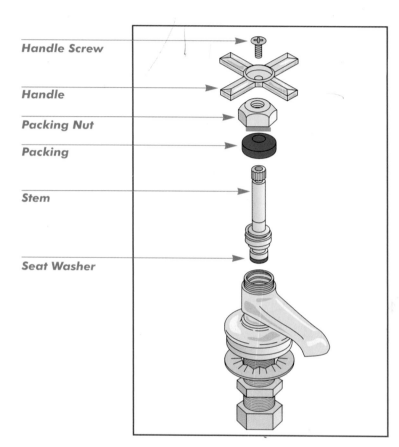

Handle Screw

Handle

Packing Nut

Packing

Stem

Seat Washer

1. OPENING UP THE FAUCET

● To stop leaks around the handle, first try tightening the packing nut. Use an adjustable wrench with the jaws taped to protect the chrome finish.

● If the leak persists, turn off the water supply, then pry off the trim cap (if any) to reach the handle screw. Pry carefully with a knife or small screwdriver *(right)*.

● Remove the handle screw *(inset)*, and pull off the handle.

● If a bent stem appears to be causing the leak, try to straighten it with a pair of tape-covered pliers. If the stem is badly damaged, replace the entire faucet *(pages 28 and 31)*.

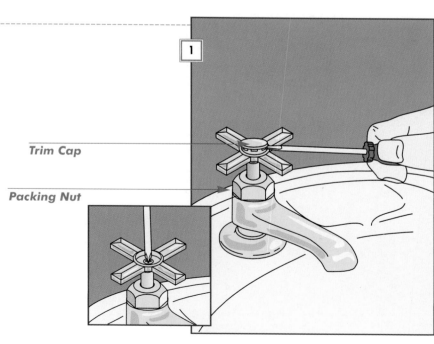

Trim Cap

Packing Nut

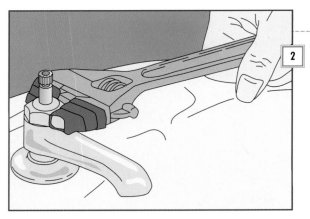

2. REMOVING THE OLD PACKING

● Unscrew the packing nut from the faucet body *(left)*; tape the wrench to prevent scratches.

● Pry off the packing washer or unwind the packing string. Be sure to remove all remnants of the old packing.

● Remove mineral deposits with steel wool.

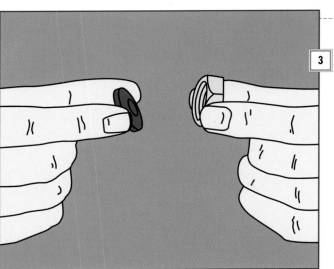

3. CHANGING THE PACKING

● Insert a new packing washer into the packing nut (*left*) or wrap new packing string around the stem several times before replacing the stem and threading the packing nut back on. Do not overtighten. Reassemble the handle and test the repair.

● Some kitchen faucets also have an O-ring or packing at the base of the spout. After lifting off the spout, remove the old packing or O-ring, replace it with new material, and reinstall the spout.

● If the spout drips, replace the seat washer (*page 11, Step 3*).

RECLAIMING OLD FIXTURES

Spend some time hunting around a plumbing salvage yard or recycling center and you may be surprised at the bargains you'll find. Restoring old fixtures to like-new working order, however, isn't always easy, so shop with a critical eye as well as a willingness to take chances.

When evaluating faucet sets with porcelain handles, look carefully for deep cracks; even small ones will weaken the handle. On the other hand, surface scratches can be touched up with some appliance or auto repair paint. Minor stains can be removed by soaking handles in chlorine bleach for several hours. Rust can be removed with a rust remover.

Don't be fooled by a sparkling chrome finish—it's what's underneath that's important. Brass is best (a solid brass faucet will be heavier than a plated steel faucet). Above all, make sure the distance between the faucet's tailpieces matches the mounting holes on your sink.

Reverse-Compression Faucets

ANATOMY

When the faucet is turned on, the spindle moves downward to create a space between the washer and the seat, allowing water to flow. (This is the opposite of the way a compression faucet works.) The seat is not designed to be removed in this type of faucet because it is subjected to little wear.

If the spout leaks, remove and replace the seat washer, or replace the entire stem assembly. To stop water from leaking around the handle, remove and replace the washer under the packing nut.

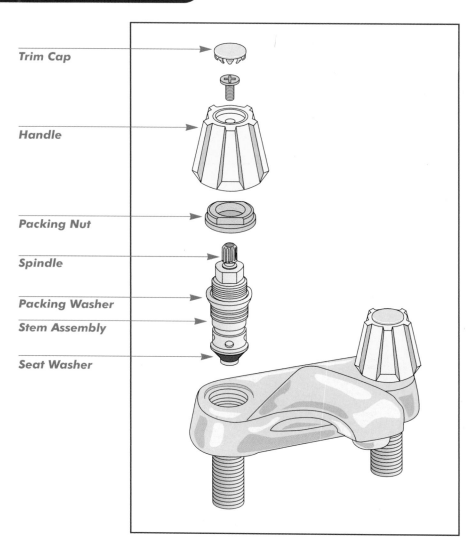

Trim Cap

Handle

Packing Nut

Spindle

Packing Washer

Stem Assembly

Seat Washer

1. OPENING UP THE FAUCET

• Turn off the water supply.

• Open the faucet handle to relieve pressure, then close the faucet.

• Carefully pry off the trim cap with a knife or small screwdriver.

• Remove the handle screw.

• Unscrew the packing nut *(right)* and lift out the stem assembly.

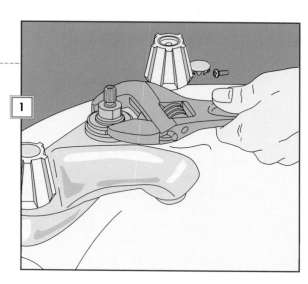

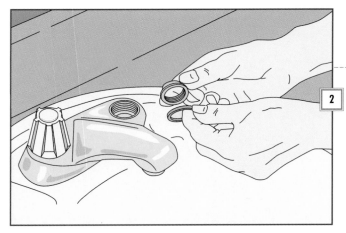

2. CHANGING THE PACKING WASHER

- To stop water from leaking around the handle, pry out the washer under the packing nut and replace it.

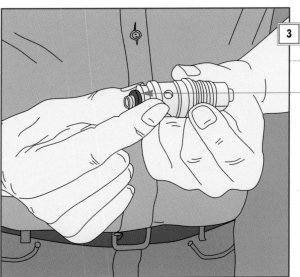

Washer

3. REPLACING THE SEAT WASHER

- Remove the seat washer *(left)* by turning the spindle by hand (or with pliers) to release the cap.

- Remove and replace the washer.

- Rethread the spindle and washer back onto the stem, then reassemble the faucet.

Ron's **TRADE SECRETS**

GETTING THE RIGHT PARTS
I've repaired hundreds of faucets in my time, and it never ceases to amaze me how many different models there are. One result is that it takes time to find the exact replacement parts you need. If I'm replacing the washers and O-rings on an old faucet, I always try to take the stem with me when I shop for parts so I can test different sizes until I get just the right fit. I've found that a local hardware store is quite likely to have the parts I need because the proprietors have been servicing neighborhood faucets for years.

Diaphragm Faucets

ANATOMY

A diaphragm faucet is a type of compression faucet and operates much like its relatives, except that it requires less maintenance. The rubber diaphragm is located on the bottom of the stem, and its edges are in contact with the walls of the valve body, sealing the stem from water. When the handle is turned to shut off the water, the diaphragm is pressed against the seat inside the valve body to seal off the water flow. Some models, such as the one at right, have a sleeve that supports the handle.

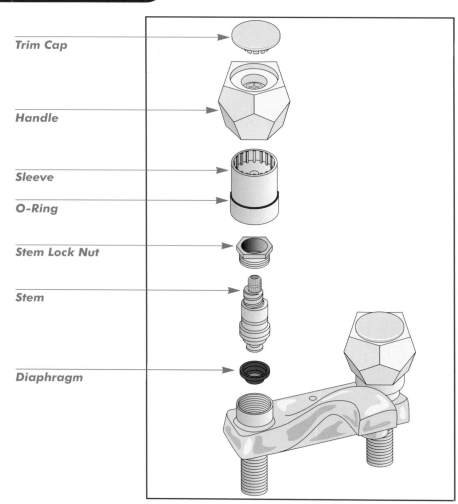

Trim Cap

Handle

Sleeve

O-Ring

Stem Lock Nut

Stem

Diaphragm

1. OPENING UP THE FAUCET

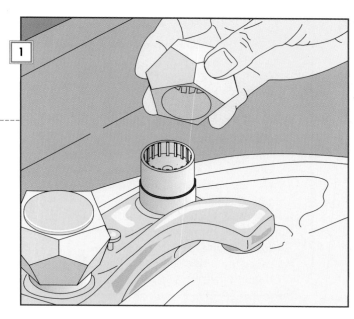

• Turn off the water supply.

• Open the faucet one half-turn, then close it. Close the drain to prevent loss of parts.

• Taking care not to mar the handle, pry off the trim cap with a knife or with a small screwdriver.

• Remove the handle screw and pull off the handle *(right)*.

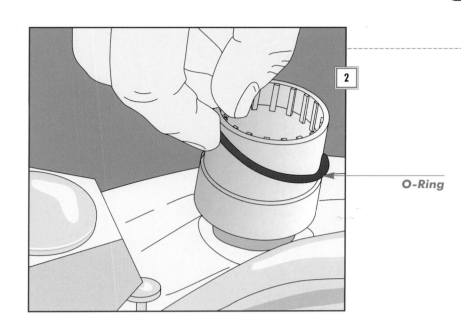

O-Ring

2. REPLACING THE O-RING

● To fix a loose handle, remove the stiff or damaged O-ring by rolling it off the sleeve. Roll the new O-ring into place.

● Reinstall the handle.

● If the spout leaks, continue disassembly in order to replace the diaphragm *(Steps 3 and 4).*

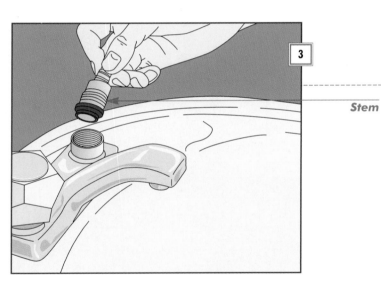

Stem

3. GETTING AT THE DIAPHRAGM

● If your faucet has a sleeve, lift it off by hand, or use pliers if necessary.

● Unscrew the stem lock nut *(anatomy)* with an adjustable wrench.

● Lift the stem from the body *(left).*

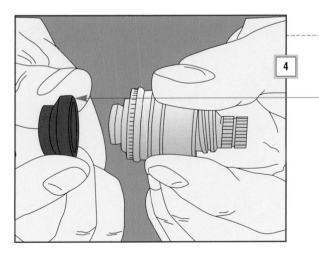

Diaphragm

4. REPLACING THE DIAPHRAGM

● Pry off the hat-shaped diaphragm by hand *(left)*, then press an exact replacement in its place.

● Insert the stem in the faucet, then tighten the lock nut firmly with a wrench.

● Reinstall the handle and test the faucet.

Disk Faucets

ANATOMY

In a disk-type faucet, a disk assembly takes the place of a stem. As the faucet is opened, the assembly rises and breaks contact with the spring-loaded seat, allowing the water to flow. An O-ring prevents water from leaking around the handle.

A defective disk assembly may cause a leak from the handle or spout. If the assembly is not damaged, a spout leak is the result of a faulty seal; a handle leak points to a worn O-ring. You can replace parts one at a time, or you can replace all of them simultaneously to avoid having to disassemble the faucet again soon.

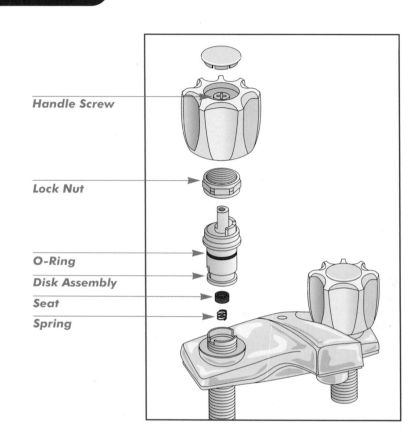

Handle Screw

Lock Nut

O-Ring
Disk Assembly
Seat
Spring

1. LIFTING OUT THE DISK ASSEMBLY

• Turn off the water and open the faucet one half-turn, then close the faucet.

• Close the drain to prevent loss of parts.

• Taking care not to mar the handle, pry off the trim cap with a knife or small screwdriver.

• Remove the handle screw and pull off the handle.

• Unscrew the lock nut with an adjustable wrench, then lift the disk assembly from the faucet body *(right)*.

• If the disk assembly is not cracked or pitted, proceed to Steps 2 and 3. Otherwise, buy a new assembly and install it, lining up its slots with those on the faucet body. Tighten the lock nut firmly.

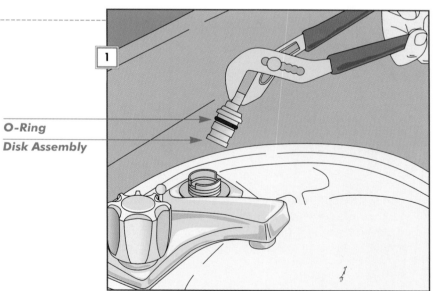

O-Ring
Disk Assembly

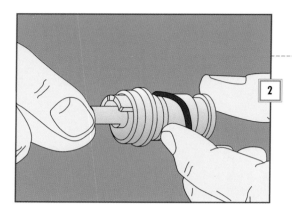

2. REPLACING THE O-RING

- If the disk assembly looks fine and only the O-ring is worn, pinch off the O-ring *(left)* and replace it.

- Reassemble and test the faucet.

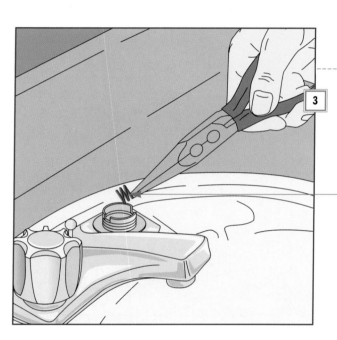

Metal Spring

3. SERVICING THE SEAT AND SPRING

- Buy a spring-and-seat repair kit for the same make and model of faucet.

- Pick the rubber seat and the spring out of the faucet body with long-nose pliers *(left)*. Note the orientation of the spring for correct reassembly; some springs are tapered.

- Replace these parts with new ones from the repair kit.

- Insert the disk assembly, lining up its slots with the faucet body.

- Tighten the lock nut, reinstall the handle, and test the faucet.

OTHER DISK FAUCETS

Some models of disk-type faucets have a ceramic seal and O-ring instead of the rubber seat and spring. Make certain that you have a repair kit intended for your specific make and model of faucet. If you aren't sure, take the whole assembly to a plumbing supplier and ask for help on getting the right parts.

Single-Lever Faucets (Rotating-ball)

ANATOMY

A rotating-ball faucet relies on the position of the ball to control water flow and temperature. With the faucet off, the ball depresses spring-loaded seats, closing them. When the faucet is turned on, the ball lifts and allows water to flow through the seats and out the faucet. Pushing the handle to the left allows more hot water to flow through the ports in the ball; pushing it to the right allows more cold water to flow.

The faucet has no washers, though it does have various other parts that must eventually be replaced. When it begins to leak from the handle, you may be able to fix it with Steps 1 and 2 on the facing page. If the leak persists, the best remedy is to rebuild the faucet. Repair kits are available for this purpose.

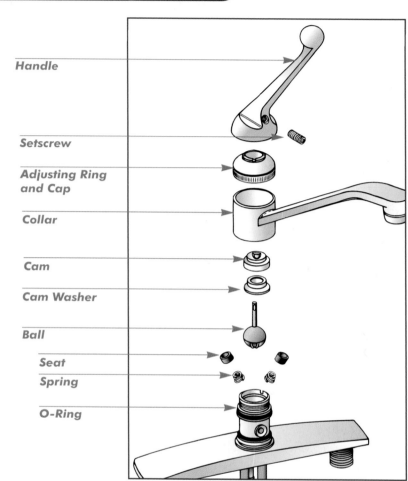

Handle

Setscrew

Adjusting Ring and Cap

Collar

Cam

Cam Washer

Ball

Seat

Spring

O-Ring

TRADE SECRETS

KEEPING TRACK OF SMALL PARTS
When I start taking apart a faucet assembly, or anything else that has a bunch of small parts, I never trust my memory. I like to keep an egg carton or two handy for such occasions. As I remove each part, I place it in a hole. When I'm ready to reassemble the piece, I just replace the parts in reverse order. Also, I find that I lose fewer parts by having a special place to store them. If you don't have any egg cartons around, a couple of old muffin tins or small plastic bags can work just as well.

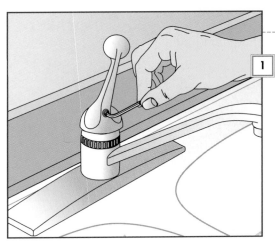

1. REMOVING THE HANDLE

- Loosen the setscrew under the handle using a small hex wrench *(left)*.

- Lift the handle from the ball lever to expose the adjusting ring.

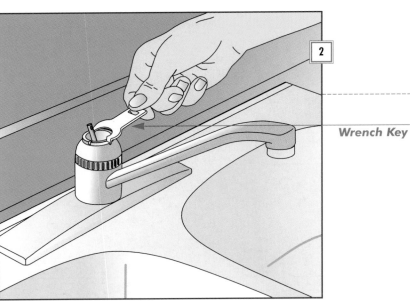

Wrench Key

2. TIGHTENING THE ADJUSTING RING

- Turn the adjusting ring clockwise. Use the edge of an old dinner knife, or a special wrench *(left)* included in the repair kit. Do not overtighten; the ball should move easily without the handle attached.

- Reinstall the handle, aligning the setscrew with the flat spot on the ball lever, and test the faucet. If the leak persists, go to Step 3.

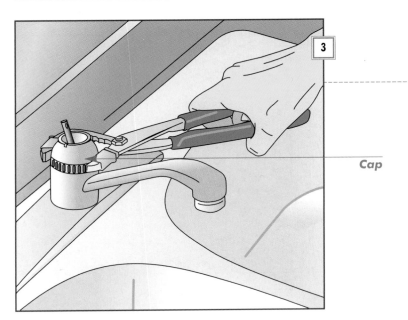

Cap

3. REMOVING THE CAP

- Turn off the water supply and open the faucet. Close the drain to prevent loss of small parts.

- Unscrew the adjusting ring and cap *(left)* by hand or with a pair of channel-joint pliers (taped to protect chrome parts).

4. REMOVING THE CAM ASSEMBLY

- Lift off the plastic cam *(right)*, exposing the cam washer and rotating ball.

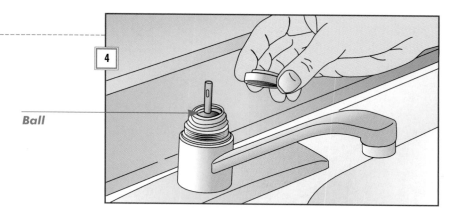

Ball

5. REPLACING THE SEATS, SPRINGS, AND BALL

- Lift the rotating ball from the faucet body *(right)*. Be sure to note its position for correct reassembly later.

- With long-nose pliers or the end of a screwdriver, remove the rubber seats and metal springs *(inset)*--or the ceramic seats and O-rings, if your faucet has them.

- Replace the old parts with new ones from the repair kit, making sure they are properly seated in the faucet body before reassembling the faucet.

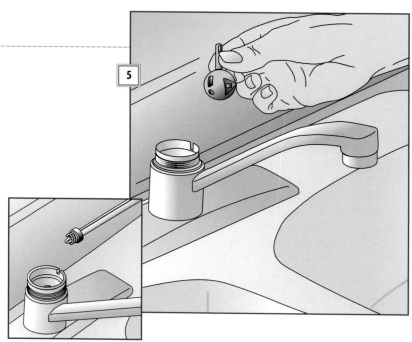

6. REPLACING THE SPOUT O-RINGS

- Twist the spout off the faucet body.

- Pry off the existing O-rings with a small screwdriver.

- Lubricate the new O-rings with petroleum jelly, then roll them into place.

- Lower the spout straight down over the body *(right)* and rotate it until it rests on the plastic slip ring at the base.

- Reassemble and test the faucet.

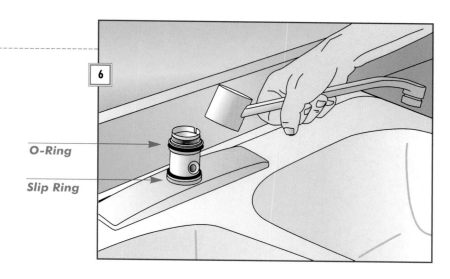

O-Ring

Slip Ring

Single-Lever Faucets (Cartridge)

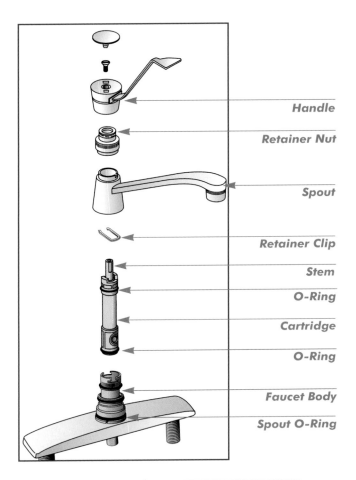

Handle

Retainer Nut

Spout

Retainer Clip

Stem

O-Ring

Cartridge

O-Ring

Faucet Body

Spout O-Ring

ANATOMY

Though its workings differ considerably from those of a rotating-ball faucet, a cartridge faucet operates under a similar principle: Its position controls water flow. As the stem is moved up and down by the lever handle, it admits more or less water. Water temperature is controlled by the rotation of the cartridge. The cartridge O-rings create a watertight seal between the cartridge and the faucet body.

To stop a leaking handle or a dripping spout, replace the O-rings or the entire cartridge *(Step 4)*. To stop leaks from the spout collar, replace the spout O-rings *(Step 6)*.

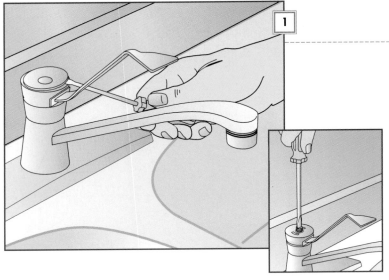

1. PRYING OFF THE TRIM CAP

• Turn off the water supply, then raise and lower the handle several times to empty the faucet.

• Close the drain to prevent loss of small parts.

• Taking care not to scratch the chrome finish, pry off the trim cap with a small screwdriver *(left)* or knife.

• Remove the handle screw *(inset)*.

2. REMOVING THE HANDLE

• The faucet handle is held in place by a lip that secures it to the retainer nut. Tilt the handle lever up sharply to unhook it from the nut.

• Lift the handle free *(right)*.

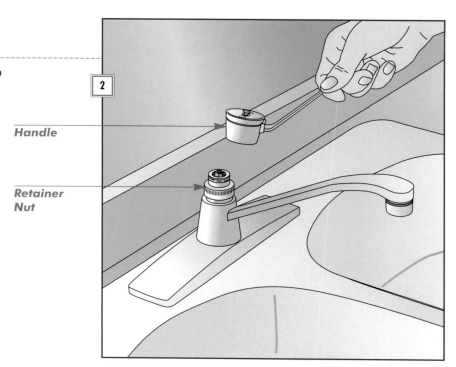

Handle

Retainer Nut

3. REMOVING THE RETAINER NUT

• Unscrew the retainer nut with channel-joint pliers, and lift off the faucet body *(right)*. If the nut will be partially visible after the faucet handle is put back in place, tape the jaws of the pliers first to prevent scratching the nut.

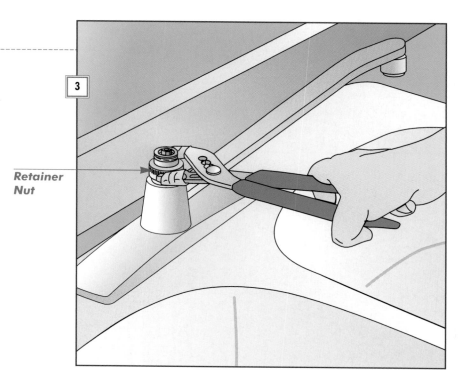

Retainer Nut

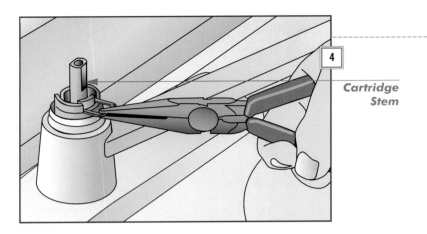

Cartridge
Stem

4. FREEING THE CARTRIDGE

● Locate the U-shaped retainer clip that holds the cartridge in the faucet body. Only the base will be visible.

● Pull the clip from its slot using long-nose pliers or tweezers *(left)*. The clip is small and easily lost; put it in a safe place until you need it for reassembly.

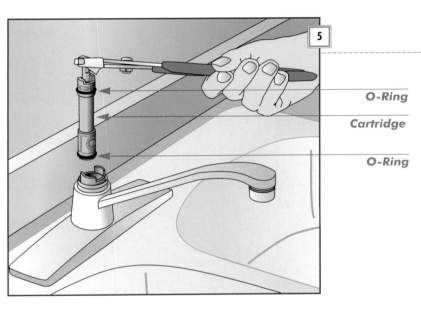

O-Ring

Cartridge

O-Ring

5. SERVICING THE CARTRIDGE

● Using taped pliers to avoid scratching the stem, grasp the cartridge stem and lift it out of the faucet body *(left)*.

● If the cartridge is worn or damaged, re-place it, being sure to align its ears with the slots on the faucet body. If only the O-rings are worn or cracked, replace them instead. Make sure that the new O-rings rest in the appropriate grooves.

● Replace the cartridge and retainer nut.

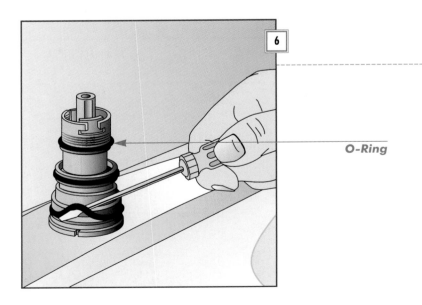

O-Ring

6. REPLACING THE SPOUT O-RINGS

● Lift the spout from the faucet body and pry off the cracked or worn O-rings *(left)*.

● Lubricate new O-rings with petroleum jelly, then roll them into the appropriate grooves.

● Slip the spout back onto the faucet body, then reassemble the faucet.

● Test the faucet. If the hot and cold water are reversed, rotate the cartridge stem one half-turn.

Single-Lever Faucets (Ceramic disk)

ANATOMY

A ceramic-disk faucet has two disks. The smaller one is on the underside of the cartridge, often connected to the faucet body by an adapter; it contacts the lower, larger disk located inside the faucet body. When the lever is in the lowered position, the upper disk is positioned at the rear of the lower disk and covers all the inlet ports, preventing water flow. As the lever is raised, it causes the upper disk to slide forward, uncovering the hot- and cold-water inlet ports as it moves. Pushing the lever side to side uncovers more or less of each port, resulting in warmer or cooler water flow.

If dirt or debris gets between the two disks, it can reduce their ability to seal the inlet ports. Other leakage problems may be remedied either by replacing the seal on each port, or by replacing the entire cartridge.

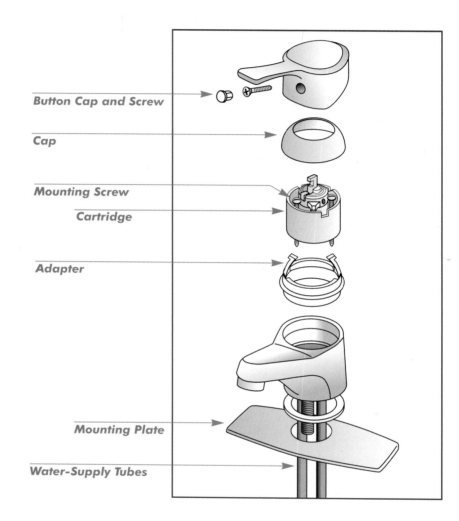

Button Cap and Screw

Cap

Mounting Screw

Cartridge

Adapter

Mounting Plate

Water-Supply Tubes

1. LOOSENING THE SETSCREW

• Turn off the water supply, then drain the faucet by lifting the lever to its highest position.

• Close the sink drain to prevent the loss of small parts.

• Pry off the button cap, if you find one, with a knife or small screwdriver, then loosen the handle screw *(right)*.

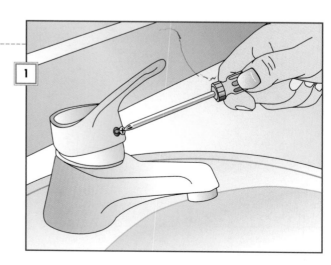

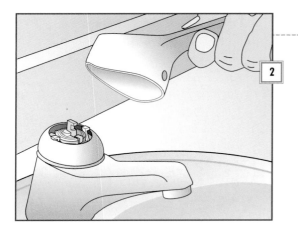

2. LIFTING OFF THE LEVER

• Lift off the handle *(left)*, exposing the cap and cartridge inside the faucet body.

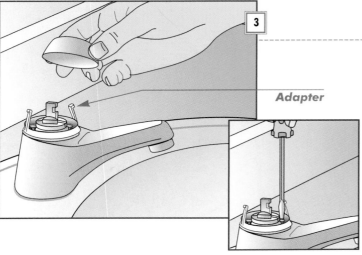

Adapter

3. FREEING THE CARTRIDGE

• Remove the cartridge cap. On some faucets, gently pry it loose from the adapter *(left)*; on others, unscrew it from the faucet body.

• Loosen the screws or retainer holding the cartridge to the faucet body *(inset)*.

• Lift out the cartridge.

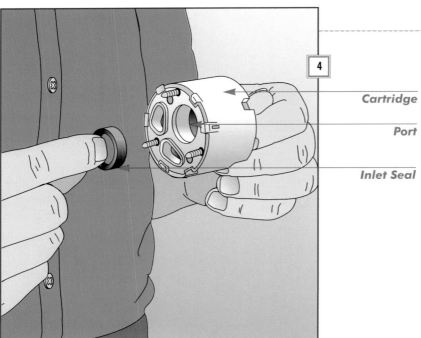

Cartridge

Port

Inlet Seal

4. REPAIRING THE CARTRIDGE

• First check to be sure that the leak is not caused by a piece of dirt between the ceramic disks. Clean the inlet ports and the surface of the lower disk inside the faucet.

• If the upper disk is cracked or pitted, buy a replacement cartridge for the same make and model of faucet. If the old seals are faulty, replace them with new ones *(left)*.

• Position the cartridge in the faucet body. Check that the ports on the bottom of the cartridge align with those of the faucet body. Screw the cartridge into place.

• Reinstall the cartridge cap and handle, then test the faucet.

Replacing a Faucet Set (Rigid supply-lines)

1. LOOSENING THE SUPPLY TUBES

If the putty beneath the faucet is dry and cracked, that may be the source of leaks between the sink and faucet body. If the underside of the faucet body is pitted badly, replace the faucet set.

● Close the shutoff valves and open the faucet to relieve pressure inside it.

● Loosen the coupling nuts at the shutoff valves with a taped adjustable wrench *(right)*, then unscrew them by hand.

Coupling Nut

Supply Tube

Coupling Nut

2. REACHING THE BASIN COUPLING NUTS

● Unscrew the basin coupling nuts under the sink *(right)*. Use a basin wrench *(inset)* if space is cramped.

Basin Wrench

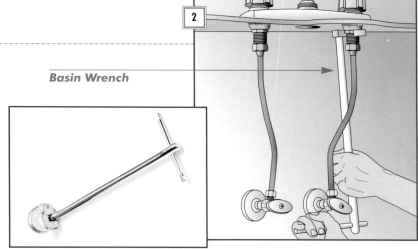

3. REMOVING THE SUPPLY TUBES

● Once you have unscrewed all the coupling nuts, pull the supply tubes out *(right)*; you may have to bend them slightly to do so, but be careful not to kink them.

● Use the basin wrench to loosen the lock nuts on each tailpiece, then remove the lock nuts by hand.

Lock Nut

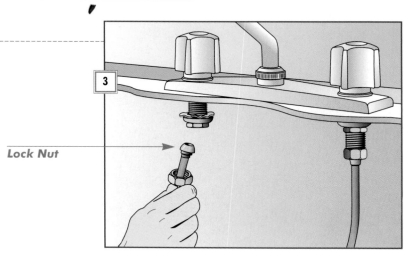

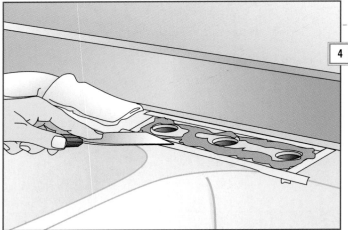

4. LIFTING OFF THE FAUCET BODY

- Lift off the faucet set.

- Protect the sink with masking tape, then scrape away the old putty with a putty knife *(left)*.

- Scour off any remaining putty residue using fine steel wool.

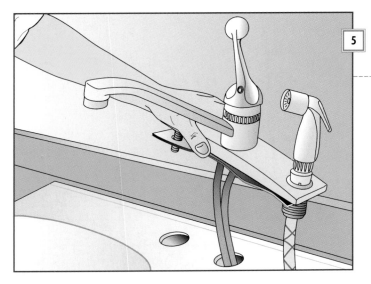

5. DROPPING IN THE NEW FAUCET

- Single-lever faucets are widely available for three-hole bathroom sinks with a 4-inch space between tailpiece centers, and kitchen sinks with a 6- or 8-inch space.

- Feed the spray hose, if any, through its hole, then work the faucet's supply tubes through the center hole of the sink *(left)*.

- Bed the faucet's mounting plate in a continuous rope of plumber's putty and push the plate into position. (Many new faucets come with deck gaskets and do not require putty. If this is the case, position the gasket beneath the mounting plate before you feed the spray hose and supply tubes through the sink's holes.)

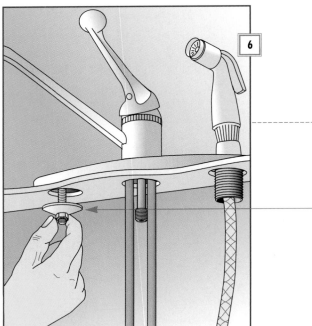

Flange

6. TIGHTENING THE SINK CONNECTIONS

- Slip the flange onto the faucet mounting bolt beneath the sink *(left)*.

- Thread a lock nut onto the bolt and tighten it with a basin wrench (tighten a plastic lock nut by hand).

- Thread a second lock nut on the sprayer hose tailpiece and tighten in the same way.

- Scrape away excess putty from around the mounting plate.

7. BENDING THE SUPPLY TUBES

• To prevent kinks, bend the copper supply tubes by hand, or with an inexpensive coiled pipe bender (available at home centers).

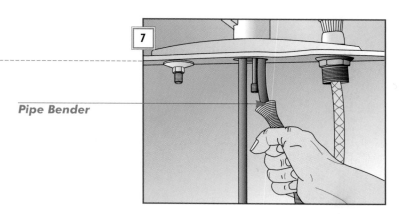

Pipe Bender

8. CONNECTING THE TUBES

• Fit a coupling nut and compression ring onto the free end of a supply tube.

• Slide the nut and ring out of the way, then push the end of the tube into the shutoff valve as far as it will go *(right)*.

• Tighten the nut by hand, then give it a quarter-turn with an adjustable wrench. Attach the other supply tube in the same way.

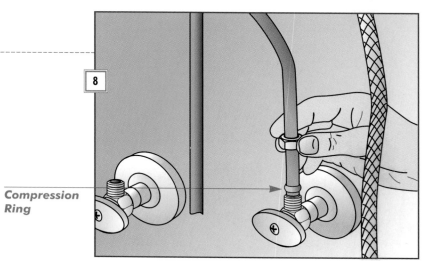

Compression Ring

9. CONNECTING THE HOSE

• To attach a spray hose that is part of the faucet set, screw its coupling nut onto the stub-out behind the supply tubes *(right)*.

• Tighten the nut with a basin wrench.

• To check the installation, unscrew the aerator on the faucet and on the sprayer *(page 6)*. Turn on the water, slowly at first, and run it alternately through the faucet and the sprayer. If leaks occur, tighten the coupling nuts another quarter-turn, but take care not to overtighten them.

• Run water full force to flush the lines.

• Replace the aerators.

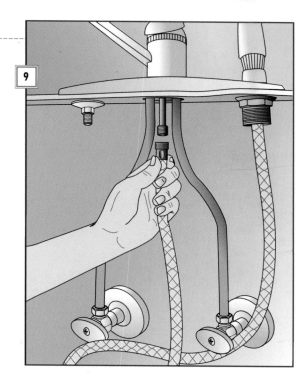

Replacing a Faucet Set (Flexible supply-lines)

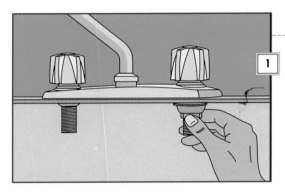

1. INSTALLING THE FAUCET SET

Faucets with threaded tailpieces can be hooked to any type of flexible supply tube.

● Remove the old faucet set *(page 28)* and clean the old putty from the sink.

● Install the faucet set as shown in Steps 5 and 6 on page 29. The new faucet set, however, won't have rigid supply tubes.

● Thread a lock nut and mounting flange (if any) onto each tailpiece *(left)* and tighten them with a basin wrench; tighten plastic lock nuts by hand.

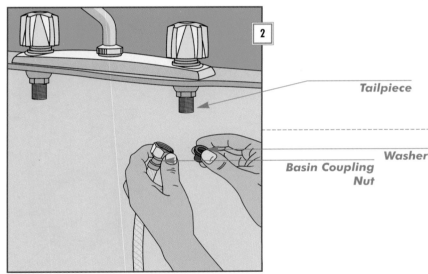

Tailpiece

Washer

Basin Coupling Nut

2. ATTACHING THE UPPER END OF THE SUPPLY TUBES

● Place one of the washers that came with the supply tube inside the large coupling nut at one end *(left)*. Some supply tubes come with washers already installed.

● Screw the nut onto the faucet tailpiece or threaded adapter and tighten it with a basin wrench.

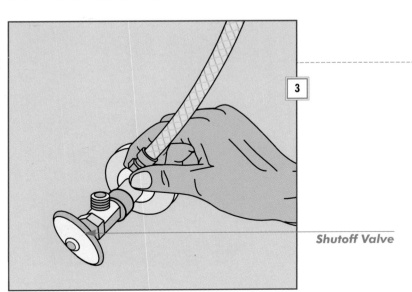

Shutoff Valve

3. ATTACHING THE LOWER END OF THE SUPPLY TUBES

● Install another washer inside the small coupling nut, then screw the nut to the shutoff valve outlet *(left)*. Tighten the nut by hand, then give it a quarter-turn with a taped adjustable wrench. Repeat with the other supply tube.

● Turn on the shutoff valves. If there are any leaks, tighten the affected coupling nut another quarter-turn.

● Remove the aerator and run water, slowly at first, to flush the lines before replacing it.

Sink Sprays and Aerators

ANATOMY

Sink spray systems connect to the faucet at a diverter valve, which is often contained within the faucet body. When the faucet is turned on, water in the spray hose keeps the diverter closed and all the water flows instead through the spout. When the spray lever is depressed, back pressure is released and the diverter valve pops open, allowing water to flow through the spray head. This pressure differential is easily disturbed by sand, rust, or grit in the aerators, in the diverter valve, or in the spray shutoff valve.

The aerator is found on faucets and sprayers. It mixes air with water to provide a "softer" flow that is more splash-free than an unaerated flow. Like spray nozzles, aerators are prone to disruption by contaminants. When reassembling an aerator, take care to keep the various seals, screens, and disks in exactly the same order as you removed them.

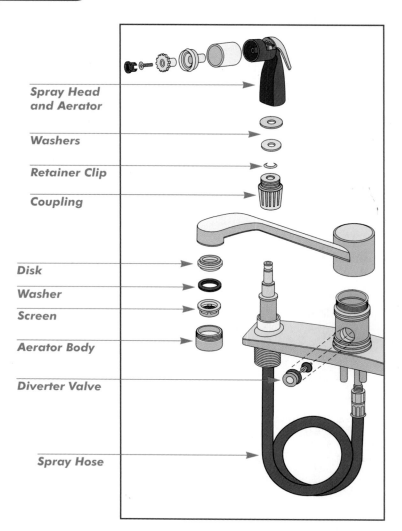

Spray Head and Aerator

Washers

Retainer Clip

Coupling

Disk

Washer

Screen

Aerator Body

Diverter Valve

Spray Hose

FAUCETS WITH INTEGRAL SPRAYERS

One of the best innovations in kitchen plumbing has been the introduction of faucets in which sprayer and spout are combined. The spray head sits in the faucet body for normal use, but pulls out to function as a conventional sprayer. With this arrangement, the sink does not require a separate hole for the sprayer. Repairing these units calls for the same techniques used on standard faucets and sprayers. As with any kind of faucet, quality varies from one model to another. The best have brass bodies and cartridges. Lesser faucets use more plastic and, although lighter in weight, are not as durable.

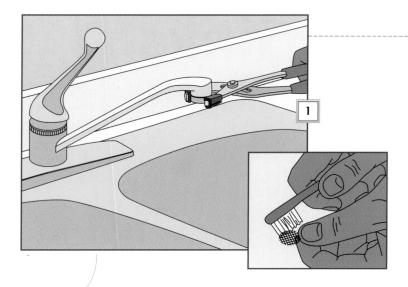

1. CLEANING THE AERATOR

● Using tape-wrapped pliers, unscrew the aerator from the spout *(left)*.

● Disassemble the aerator, taking care to note the exact order of parts.

● Examine the aerator's washer and replace it if it is worn or cracked.

● Soak the screen and aerator disks in vinegar and scrub them with a small brush *(inset)*. Turn on the faucet to flush sediment from the spout.

● Reassemble the aerator and thread the assembly back onto the spout. Tighten it one-quarter turn with the pliers.

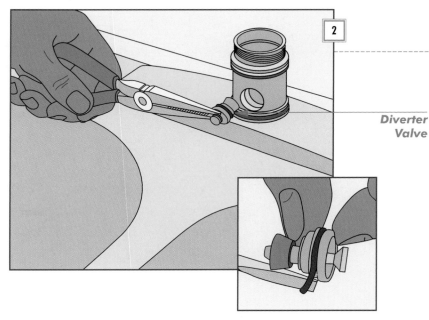

Diverter Valve

2. SERVICING THE DIVERTER VALVE

● Turn off the water supply to the faucet, then remove the spout *(pages 22 and 25)*.

● For a single-lever faucet, pull out or unscrew the valve. For a double-handle faucet, unscrew the valve under the spout nut.

● If the valve is bent or the conical washer *(inset)* is loose, buy a new valve.

● Otherwise, rinse off the valve, pry off the O-ring *(inset)*, and roll a replacement onto the valve. Then flush the diverter socket with water.

● Insert the valve, reassemble the faucet, and turn on the water. If leaks persist, clean the spray head *(below)*.

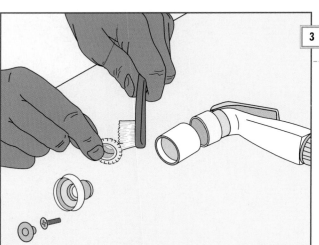

3. CLEANING THE SPRAY HEAD

● Pry the screw cover from the nozzle with a knife or small screwdriver.

● Remove the screw to free the perforated disk, seat, and sleeve. Set these parts aside in exact order of disassembly. Buy replacements for any washers.

● To remove mineral deposits, soak the disk and seat in vinegar, then scrub with a small brush *(left)*. Reassemble the sprayer.

FIX IT: Kitchen Sinks

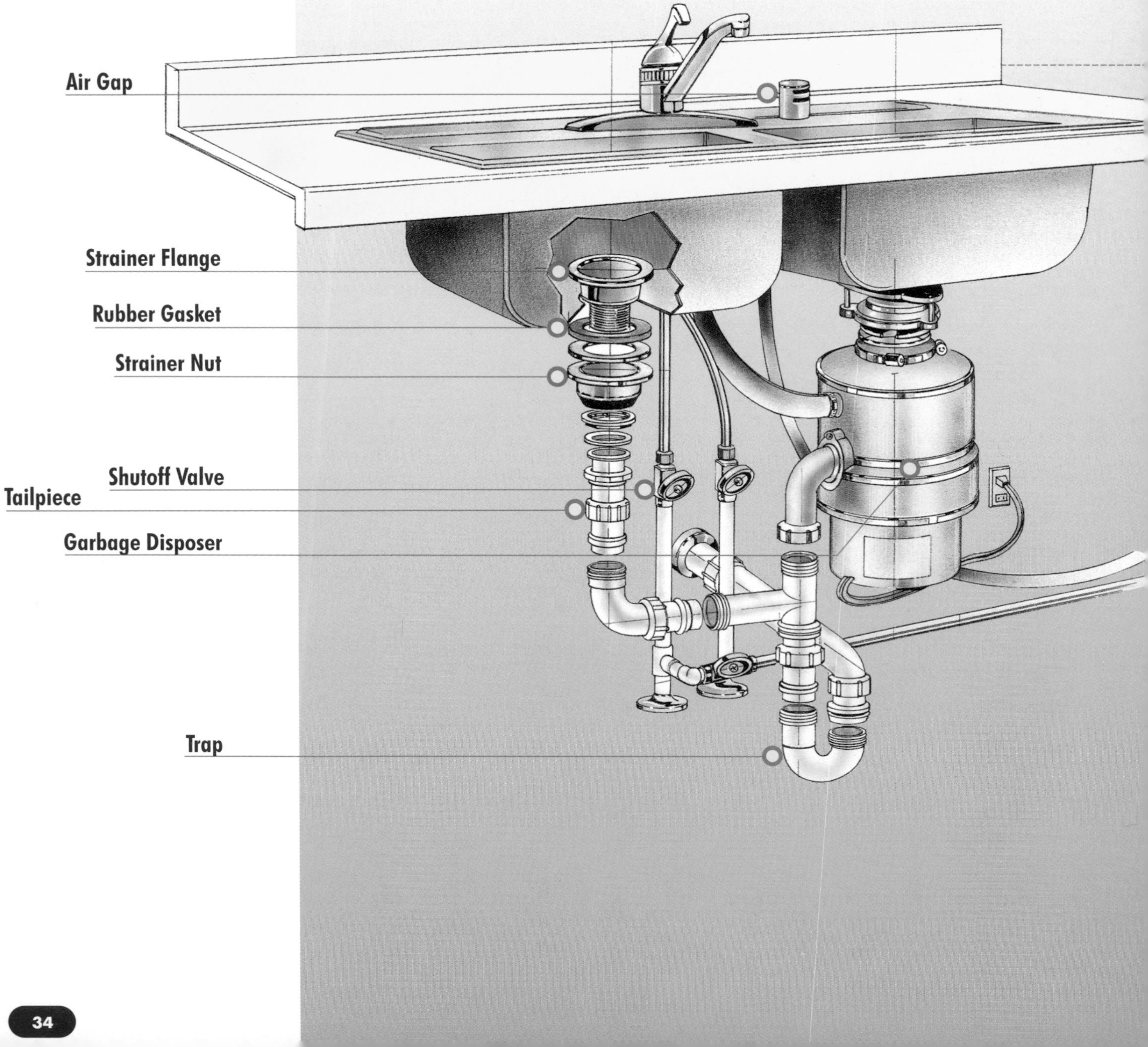

Air Gap

Strainer Flange

Rubber Gasket

Strainer Nut

Shutoff Valve

Tailpiece

Garbage Disposer

Trap

Chapter 2

Contents

How They Work

Many modern kitchen sinks are equipped with a garbage disposer, through which water drains from the sink and from the dishwasher, if there is one. In double-sink installations like the one shown at left, a drainpipe for the second sink, comprising a number of pipes and fittings, joins the garbage disposer drainpipe at an intersection called a continuous waste T. From there, waste water passes through a trap, which blocks sewer gas from entering the house, and into a trap arm leading to the drainpipe in the wall. Some plumbing codes require that a dishwasher be fitted with an air gap--a simple device mounted next to the faucet that prevents back-siphonage of waste water into the water supply.

As explained on page 37, any of the pipe joints under the sink may begin to leak, and drains occasionally become clogged with food--problems that usually respond to a wrench or a plunger.

Troubleshooting

Problem	Solution
• **Water leaks from faucet set**	Tighten lock nuts under faucet set **29** • Replace the faucet set **28, 31** •
• **Water leaks from trap fittings**	Tighten slip nuts on trap assembly **39** •
• **Water leaks from dishwasher hose**	Tighten or replace the hose clamp **42** • Trim or replace the hose **42** •
• **Water leaks from garbage disposer drainpipe**	Replace worn washer **43** •
• **Water leaks from drainpipes**	Replace worn washers; replace trap assembly **39** •
• **Water leaks from sink strainer**	Tighten lock nut or retainer screws **41** • Replace plumber's putty or worn parts **41** •
• **Water leaks from supply tubes or fittings**	Tighten coupling nut at shutoff valve **30** • Replace faucet that has rigid supply lines **28** • Replace flexible-type supply tubes **31** •
• **Drain blocked or sluggish**	Remove clogs in the trap bend or arm with a plunger or a hose **38** • Disassemble and clean drain fittings **39** • Use an auger to remove clogs behind the wall **40** •

Before You Start

Whether your kitchen has a single basin sink or one with a double basin, the clogs and leaks that sometimes cause trouble are often easier to fix than meets the eye.

DEALING WITH CLOGS

You can avoid most clogs altogether by placing strainer baskets in the drain openings and by pouring grease and coffee grounds into the trash, not down the drain. If a sink backs up despite your best efforts, you can try a chemical cleaner in a porcelain sink, if the drain is only partially clogged, but such chemicals may mar the finish of an enamel or stainless-steel sink. A plunger is better in these situations (*page 38*). For more stubborn clogs, you may have to disassemble the drain plumbing for cleaning, and it may be necessary to employ an auger (*pages 39-40*).

STOPPING LEAKS

Repairs for leaks from the sink strainer, the dishwasher, or the garbage disposer are shown on pages 41-43, but many other leaks dry up if you tighten the slip nuts on the drain assembly. If this doesn't work, remove the part of the trap nearest the leak, and install a new beveled washer under the connecting slip nut. However, before disconnecting any under-sink plumbing, remember to turn off the water supply to the faucet. Also, it's wise to replace the washer in every joint you disassemble before fitting it together again. On occasion, a trap may be too corroded to reinstall; replace it instead. Metal or plastic traps work equally well.

TOOLS

Plunger
C-clamp
Pipe wrench
Channel-joint pliers
Garden hose
Pan or pail
Auger
Putty knife
Screwdriver
Utility knife

MATERIALS

Plumber's putty
Beveled washers for slip nuts
Hose clamps
Pipe tape
Rubber gasket for garbage disposer

SAFETY FIRST

Before working on a dishwasher or garbage disposer, turn off the power. When using chemical drain cleaners, follow all instructions and wear safety glasses. Never use a plunger with a chemical drain cleaner--splashes are hazardous. Do not pour boiling water down plastic pipes to clear grease clogs.

Clearing a Minor Drain Clog

USING A PLUNGER

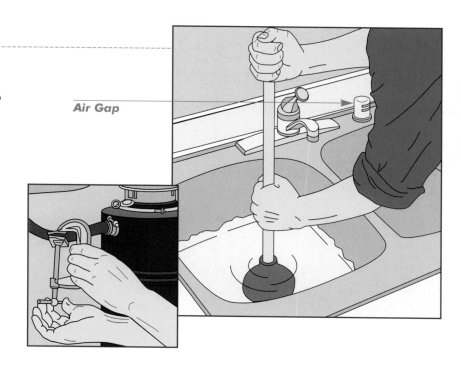

Air Gap

● In a sink with a garbage disposer, seal off the hose leading to the air gap, if there is one, by pinching it with a C-clamp and two pieces of wood *(inset)*.

● Lift out the sink basket and clear any debris caught in the drain opening.

● Fill the sink with water to cover the plunger cup. On a double sink, pack rags wrapped in plastic into the other drain opening, or hold the strainer in place.

● Set the plunger squarely over the drain and pump vigorously up and down at least a dozen times, then pull away sharply. Repeat several times, if necessary.

● If plunging does not clear the clog, use a hose *(Step 2)*.

USING A HOSE

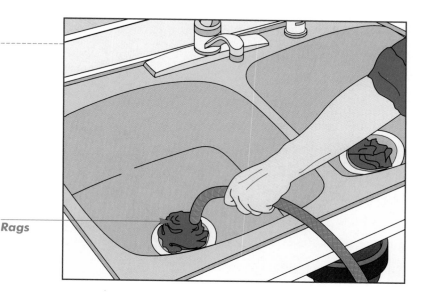

Rags

● Attach a hose to an outdoor faucet or the spout of a utility sink and bring the hose to the sink.

● Feed the hose into the drain as far as possible and pack rags tightly around it *(right)*. Alternatively, use an expansion nozzle *(page 73)*. On a double sink, pack rags wrapped in plastic into the other drain opening.

● Hold the hose firmly in place and have a helper rapidly turn the faucet on and off several times.

● If the water backs up instead of clearing the clog, you will have to remove the trap *(opposite)*.

● Because of the danger of back-siphonage, pull the hose from the drain as soon as you finish work.

Clearing a Trap and Branch Drain

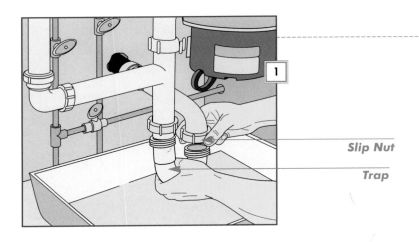

Slip Nut

Trap

1. REMOVING THE TRAP UNDER A DOUBLE SINK

• Place a pan or pail under the trap to catch drips. Support the bend with one hand and loosen the slip nuts on each end with channel-joint pliers.

• Pull the trap free *(left)* and pour out any water. Scrub the trap with a flexible brush to clear obstructions, then rinse it.

• If you removed an obstruction, reinstall the trap using new beveled washers in the slip nuts and test the drain. If it is still blocked, remove the trap and proceed to the next step.

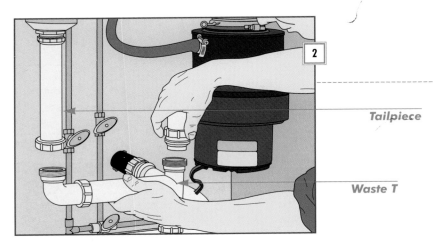

Tailpiece

Waste T

2. REMOVING OTHER TRAP FITTINGS

• Disconnect the continuous waste T by loosening the slip nuts at the garbage disposer drainpipe and at the end of the sink's tailpiece, using a wrench.

• Remove the fittings and clean them as in Step 1.

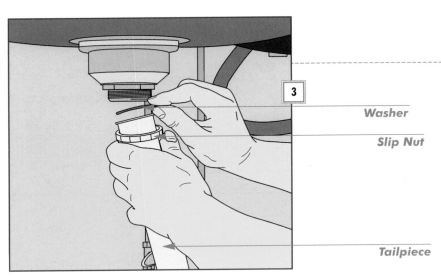

Washer

Slip Nut

Tailpiece

3. REMOVING THE TAILPIECE

• Loosen the slip nut connecting the tailpiece to the sink and remove the tailpiece. Clean the tailpiece as above.

• If the tailpiece is damaged, corroded, or if it is severely clogged, replace it with a new one. Be sure to replace any washers at the same time *(left)*.

4. REMOVING THE TRAP ARM

• If there is an escutcheon plate where the trap arm enters the wall, slide it along the trap arm to reveal the slip nut that joins the trap arm to the drain stub-out.

• Loosen the slip nut with a pipe wrench, squirting the threads beforehand with penetrating oil if the stub-out is galvanized steel.

• Unthread the slip nut by hand, then pull out the trap arm and clean it as in the previous steps.

• If you found no obstruction in the trap arm, proceed to Step 5. Otherwise, reconnect the drain assembly with new beveled washers. Coat the inside of the slip nuts with plumber's putty or silicone sealant before tightening the connections.

• Run water through the pipes and tighten any leaking connections.

• If the water still does not run freely, remove the trap and the trap arm again, and auger the branch drain *(Step 5)*.

Trap Arm

5. AUGERING THE BRANCH DRAIN

• Probe into the branch drain behind the wall with an auger. Work carefully; ramming the auger too vigorously can loosen fittings behind the wall and could damage old pipes.

• Reconnect the trap assembly using new washers, and tighten all slip nuts.

• Test the drain. If it is still clogged, it's time for professional help; call a plumber.

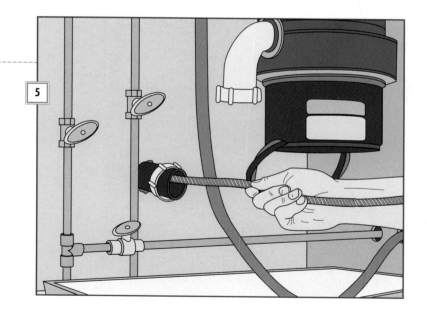

Fixing a Leaky Strainer

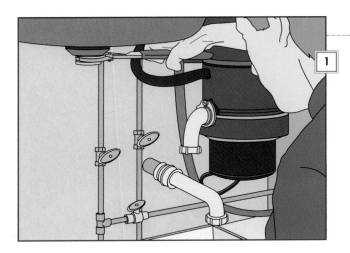

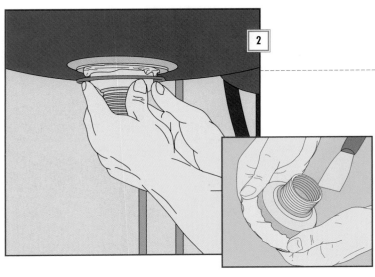

1. REMOVING THE STRAINER

Leaks develop when the strainer assembly wears through or the putty dries out.

● Remove the tailpiece. You may have to loosen or remove other trap fittings to get it off *(page 39)*.

● Unscrew the lock nut securing the strainer to the sink. If the strainer starts to turn in the sink while you loosen the lock nut, wedge a screwdriver into the drain and hold it steady with your free hand. If the strainer is held by a plastic retainer, remove its screws and twist the retainer a quarter-turn to unlock the strainer.

2. REINSTALLING THE STRAINER

● Scrape the putty from the drain hole, and from the old strainer if you plan to reuse it.

● Apply a 1/2-inch strip of plumber's putty under the lip of the strainer *(inset)*. (Some strainers come with adhesive-coated rubber gaskets and need no putty.)

● Lower the strainer body into the drain opening from above.

● From underneath the sink, slip the rubber and metal washers over the neck of the strainer *(left)*, then secure the lock nut or retainer and screws.

3. INSTALLING THE STRAINER

● Scrape away any excess putty around the drain opening with a putty knife, being careful not to scratch the surface. Reinstall the tailpiece and any other parts that you removed.

Repairing Dishwasher Connections

1. REMOVING THE DRAIN HOSE

• If water is leaking from the dishwasher drain hose between the air gap and the disposer, turn off the power to the dishwasher and disposer.

• Close the dishwasher shutoff valve.

• Loosen the clamp securing the hose to the disposer *(right)*, then work the hose off the disposer and then off the air gap.

• Replace a damaged or stiff hose with a new one.

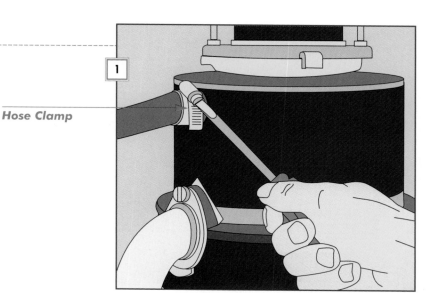

Hose Clamp

2. REFURBISHING THE DRAIN HOSE

• If only the ends are damaged, cut off an inch or two of the hose with a utility knife *(right)*.

• To reconnect the hose, or install a new one, slide new clamps onto the hose, push the ends onto the connections, and tighten the clamps.

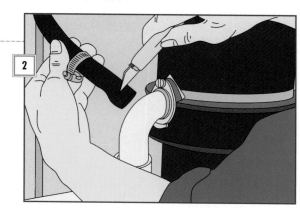

STOPPING DISHWASHER LEAKS

If the dishwasher and the disposer are near each other and the steps above fail to stop the leak, the problem may be beneath the dishwasher. Check beneath the washer for a loose connection on the water inlet line. To do this, turn off the power, remove the retaining screws in the lower front panel of the dishwasher, then lift off the panel. Tighten the water inlet line fitting with a wrench *(right)* before reinstalling the panel.

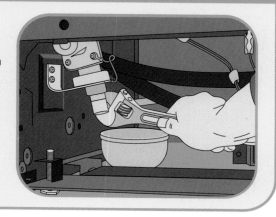

Stopping Leaks from a Garbage Disposer

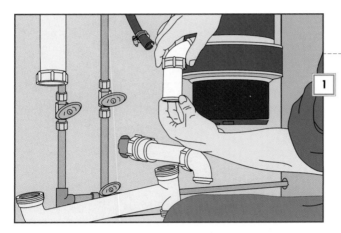

1. DISCONNECTING THE GARBAGE DISPOSER

• Turn off the water; unplug the garbage disposer and place a container under it.

• Loosen the slip nut at the end of the disposer's drainpipe. You may also have to loosen or remove various trap fittings *(left)*, before the disposer can be lowered from its mounts *(below)*. See page 39 for trap removal instructions.

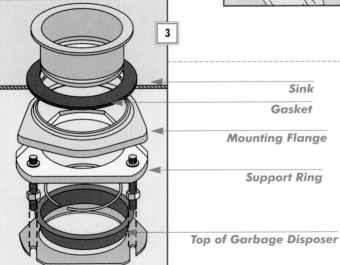

2. REMOVING THE GARBAGE DISPOSER AND THE SNAP RING

• Supporting the disposer with one hand *(left)*—it weighs several pounds— turn the lower support ring to unlock it from the mounting assembly, then lower the unit from the sink. If the support ring will not turn, use a screwdriver for better leverage.

• Remove the screws on the mounting assembly and push up the mounting flange at the base of the assembly.

• Pop the snap ring out of the strainer flange *(inset)* using a flat-tipped screwdriver. The mounting flange and gasket will come off with the snap ring.

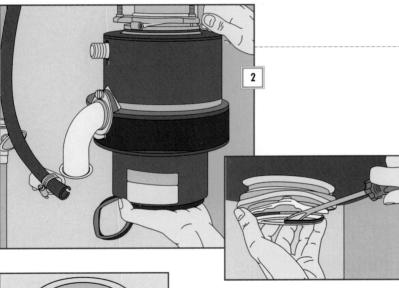

Sink

Gasket

Mounting Flange

Support Ring

Top of Garbage Disposer

3. REPLACING THE DISPOSER GASKET

• With the disposer and snap ring removed, replace the putty or gasket as described on page 41.

• Lower the strainer into the drain hole, then push a new rubber gasket *(left)* onto the strainer body from under the sink.

• Reinstall the mounting flange and snap ring and tighten the retaining screws, then lift the garbage disposer and support ring into place.

FIX IT: Bathroom Sinks

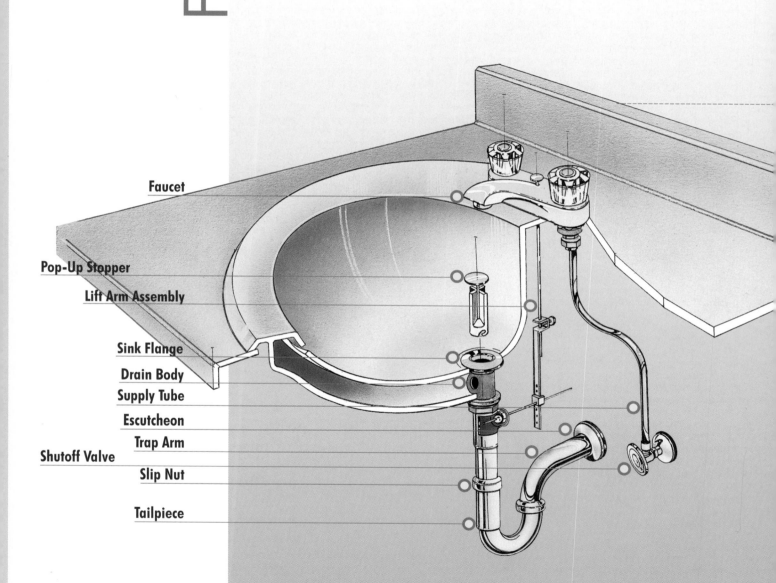

Faucet

Pop-Up Stopper

Lift Arm Assembly

Sink Flange

Drain Body

Supply Tube

Escutcheon

Trap Arm

Shutoff Valve

Slip Nut

Tailpiece

Chapter 3

Contents

How They Work

Bathroom-sink plumbing has much in common with plumbing for a kitchen sink. Supply lines feed the faucet, and waste water passes out the drain, filling a trap to prevent sewer gas from entering the house. Most traps consist of a simple, U-shaped section of pipe, as shown at left; however, if you live in a house with older plumbing, the trap may be shaped like an S. Plumbing codes often prohibit this kind of trap in new plumbing, but you can replace a faulty one with the same kind.

As in a kitchen, drains from two bathroom sinks usually meet at a continuous waste T above the trap. If both sinks empty slowly, the problem lies below the T; when only one drain is sluggish, the problem lies between the pop-up stopper, used to close the drain opening, and the T. The stopper is raised and lowered by a handle linked by a lift arm to a pivot rod to which the stopper attaches. Lifting the handle lowers the stopper to close the drain; raising the handle opens the drain.

Troubleshooting

Problem	Solution
• **Water seeps from sink**	Clean and replace O-ring on stopper **51** • Adjust stopper **51** • Adjust or replace lift mechanism **51** •
• **Pop-up stopper fails to pen or close properly**	Reconnect or readjust lift mechanism **51** • Replace drain body and pop-up assembly **52** •
• **Water leaks from faucet set**	Tighten slip nuts under faucet set **44** • Replace faucet set **28, 31** •
• **Water leaks around sink flange**	Replace worn putty or gasket **53** •
• **Water leaks around pivot rod**	Tighten retaining nut **51** • Replace pivot rod washer or gasket **51** •
• **Water leaks around trap fittings**	Tighten all connections **53** • Replace worn washers or corroded trap **49** •
• **Drain slow or blocked**	Lift out and clean stopper **51** • Adjust or replace lift mechanism **51** • Clear the drain with a plunger **48** • Remove the blockage with an auger **49** •

Before You Start

Clogs and leaks in bathroom sinks can be even easier to fix than similar problems in the kitchen, if only because bathroom-sink plumbing is less complicated.

NARROW CONDUITS

Because pipes in the bathroom are smaller in diameter than kitchen pipes, clogs are more common here. In fact, they are all but inevitable, since soapy residues may accumulate at almost any point between the pop-up stopper in the sink drain and the vertical drainpipe behind the wall that carries wastes to the sewer.

When clearing a clog, always remove the stopper from the sink. Clogs in the trap may respond to a chemical drain opener, but a plunger is safer for both you and your pipes. If plunging fails to clear the drain, you may have to disassemble the trap to clean its various sections one at a time.

Another common problem with bathroom sinks is leakage. Water under the sink usually points to worn faucet parts or loose supply or drain fittings. Replacing dried putty or tightening a slip nut may be all that's needed.

Before You StartTips:

⋯▷ When working on a sink's drainpipes, there's no need to close the shutoff valves beneath the sink.

⋯▷ Have a bucket and rags on hand to catch water runoff from trap and drain repairs.

⋯▷ To loosen corroded nuts, spray them with penetrating oil, then wait 15 minutes and try again.

TOOLS

Plunger
Auger (manual or power)
Monkey wrench
Channel-joint pliers
Pail or pan
Screwdriver
Putty knife
Adjustable wrench

MATERIALS

Chemical drain cleaner
Rags
Pipe tape
Penetrating oil
Plumber's putty

SAFETY FIRST

Wear rubber gloves and safety goggles when pouring drain cleaner; use a funnel to prevent damage to the sink and stopper.

Clearing a Minor Clog

USING A PLUNGER

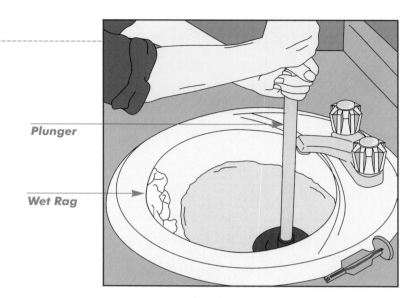

Plunger

Wet Rag

- Remove the sink stopper *(page 51)*; pack a wet rag into the sink's overflow opening.

- Coat the rim of the plunger with petroleum jelly and run just enough water in the sink to cover the plunger cup.

- Insert the plunger at an angle to avoid trapping air in the cup, then forcefully plunge down and up a dozen or more times *(right)*. Keep the plunger upright and the cup sealed over the drain.

- If the plunger is ineffective, try using an auger *(Step 2)*.

USING AN AUGER

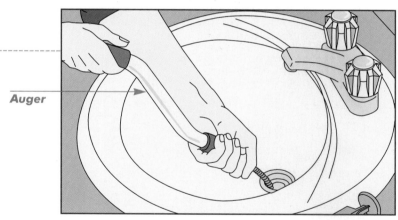

Auger

- Feed the auger into the drain as far as the bend of the trap *(right)*.

- Tighten the handle thumbscrew, then turn and push the auger to loosen the clog.

- Withdraw the auger slowly, pulling debris from the trap. If this fails, disassemble the trap *(page 49)* or use a chemical drain cleaner *(Step 3)*.

CHEMICAL DRAIN CLEANERS

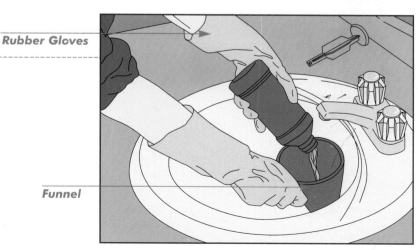

Rubber Gloves

Funnel

- Read the label carefully before buying a chemical product, as some may damage garbage disposers or plastic drainpipes.

- Wear rubber gloves and safety goggles, and pour the cleaner into a funnel to reduce dangerous splashes *(right)*.

- If you fail to remove the clog, do not try a different chemical product. A mix of chemicals can be dangerous.

Clearing a Trap and Branch Drain

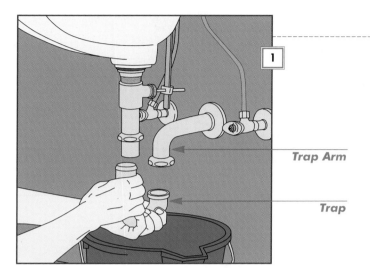

Trap Arm

Trap

1. DISCONNECTING THE TRAP

• Place a container under the trap to catch drips, and have some rags on hand. Support the trap bend with one hand and loosen the two slip nuts with a wrench, starting with the nut on the trap arm.

• Pull the trap bend down and off *(left)*, and empty the water into the container.

• If the bend is corroded, replace it. Otherwise, scrub it—a bottle brush works well—then reinstall it *(page 50, Step 5)*.

• If the clog remains, proceed to Steps 2 and 3.

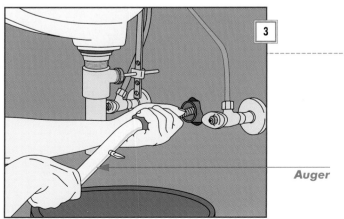

Tailpiece

Slip Nut

Escutcheon

2. REMOVING THE TRAP ARM

• After removing the trap, pry the escutcheon away from the wall and use a monkey wrench to loosen the trap-arm slip nut *(left)*.

• Unscrew the slip nut, then slide the nut and washer along the trap arm.

• Twist the trap arm free, and clean it as in Step 1. If the arm is corroded, replace it.

• You might also have to remove the tailpiece to provide room to auger *(page 53)*.

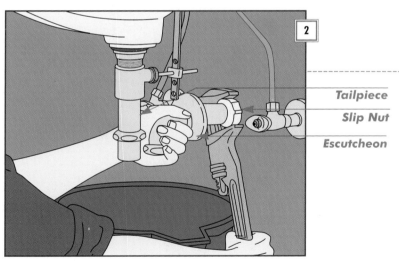

Auger

3. USING AN AUGER

• Feed the auger into the drainpipe until the tip reaches the blockage *(left)*. Lock the handle with the thumbscrew and turn the auger clockwise. As it breaks up the clog, feed in more of the auger.

• When the auger moves easily in the pipe, remove it slowly and reassemble the trap *(next page)*.

4. RECONNECTING THE TRAP ARM

• Clean away any debris at the opening in the wall.

• Slide the fittings onto the trap arm in the following order: slip nut (threaded end first), escutcheon, another slip nut (threaded end last), and beveled slip-nut washer.

• Push the trap arm about 1 1/2 inches into the drain *(right)*.

• Slide the washer onto the adapter or stub-out, then tighten the slip nut over it by hand.

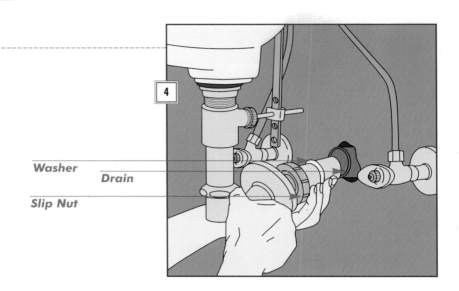

Washer

Drain

Slip Nut

5. REINSTALLING THE TRAP BEND

• Slide the slip nut onto the tailpiece, threaded end down, then slide on a washer, beveled side down.

• Push the long end of the trap onto the tailpiece and slide it up until the short end meets the trap arm.

• Gently tighten all the nuts with a wrench *(right)*. Lightly coating the slip nuts with plumber's putty or silicone sealant before assembly will help prevent leaks. If any connections leak, tighten them another quarter-turn.

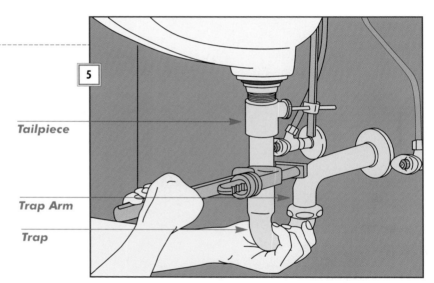

Tailpiece

Trap Arm

Trap

USING A ROTARY AUGER

A rotary auger stores the auger coil in a case between uses and aids in turning the auger to clear a drain. The model shown here is turned by hand; a variation fits an electric drill. If you are confident in working with heavy-duty power tools, consider renting a power auger. Also called a drain gun, it comes with its own motor and usually succeeds where a hand auger might fail. When using a drain gun—or even an electric-drill powered rotary auger—follow speed recommendations closely. Too much speed can damage the sink or the drain piping.

Adjusting the Pop-Up Mechanism

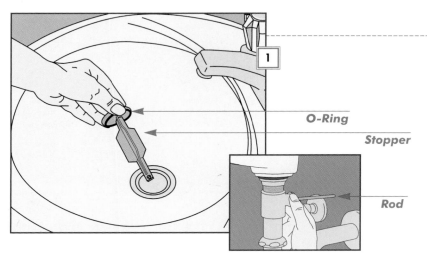

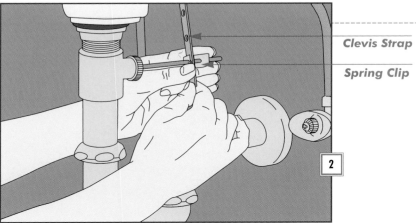

O-Ring

Stopper

Rod

Clevis Strap

Spring Clip

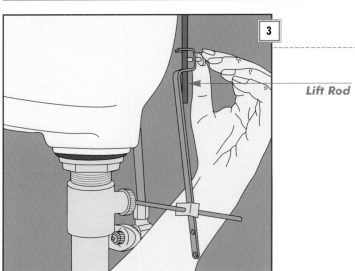

Lift Rod

1. REMOVING THE POP-UP STOPPER

• Pull the stopper from the drain *(left)*. If necessary, twist the stopper to free it from the pivot rod, or unscrew a retaining nut below the sink to loosen the rod *(inset)*.

• Clean the stopper and replace the O-ring.

• Reinsert the stopper, reconnect the pivot rod, and tighten the retaining nut.

• Fill the basin. If water leaks out, adjust the lift mechanism *(Steps 2 and 3)*.

2. ADJUSTING THE PIVOT ROD

• Pinch the metal spring clip and slip one side off the end of the pivot rod *(left)*.

• Slide the end of the pivot rod out of its hole in the clevis strap, then reinsert it.

• Reinstall the metal spring clip and test the lift assembly. Adjust the pivot rod as needed until the stopper seats properly. You may have to slip the rod into a higher or lower hole (experiment to see which works best).

3. ADJUSTING THE LIFT ROD

• If the stopper still doesn't seal the drain, adjust the lift rod. Loosen the clevis screw with pliers, then unscrew it by hand *(left)*.

• To keep water from seeping out of the sink, push the clevis strap up the lift rod to shorten the assembly.

• Tighten the clevis screw and test, re-adjusting if necessary.

• If water continues to leak out of the sink, replace the drain body *(page 52)*.

Replacing the Drain Body

1. REMOVING THE OLD DRAIN BODY

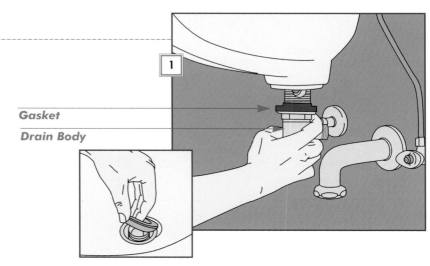

- Remove the trap *(page 49)*. Disconnect the pop-up lift mechanism *(page 51)*.

- With a pipe wrench, loosen the lock nut holding the drain body to the sink. To keep the drain body from rotating, wedge a screwdriver down the drain opening and into a slot in the drain body.

Gasket

Drain Body

- Push the drain body upwards, then un-screw or lift off the sink flange *(inset)*. Pull the drain body down and out *(right)*.

2. INSTALLING THE NEW DRAIN BODY

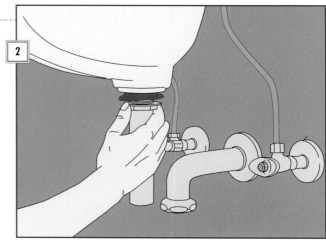

- Press a rope of plumber's putty around the drain hole.

- After removing the pivot rod from the new drain body, set it into the opening and reinstall the sink flange. Slip on the rubber gasket; snug it against the underside of the sink with the lock nut.

- Turn the drain body pivot hole to the wall, tighten the lock nut with a wrench, then reinstall the pivot rod and connect the pop-up *(page 51)*. Reinstall the trap *(page 50)* and test for leaks.

TYPES OF DRAIN BODIES

A sink's drain body contains the pop-up stopper, pivot rod, and tailpiece, along with gaskets and washers for sealing the assembly to the sink. Drain bodies may be one-piece models, as described above, or two-piece-threaded units that screw together.

Plastic drain bodies *(far right)* are inexpensive and widely available, but you may have to visit a well-stocked plumbing sup-ply store or a large home cen-ter if you prefer a drain body made of brass *(near right)*.

Stopping Leaks Under the Sink

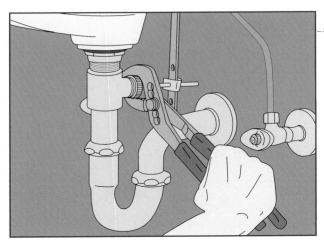

TIGHTEN THE LIFT ROD RETAINING NUT

• If water is leaking around the retaining nut, tighten the nut with pliers *(left)*.

• If leaks continue, remove the retaining nut, slide the pivot rod out, and replace the washer or gasket under the retaining nut.

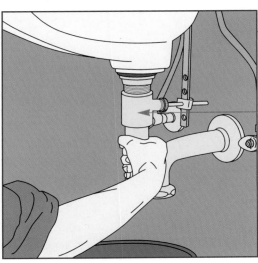

Tailpiece

TIGHTEN THE TAILPIECE

• If water is leaking around the tailpiece, loosen the slip nut at the bottom of the tailpiece.

• Tighten the tailpiece by hand *(right)*.

• Using a wrench, tighten the slip nut at the bottom of the tailpiece.

REPACK THE SINK FLANGE

• A worn washer or cracked putty under the sink flange may allow water to seep below the sink. Remove the trap *(page 50)*, and the drain body *(page 52)*, and disconnect the lift rod *(page 51)*. Free the flange from the sink.

• Press a thin rope of plumber's putty under the lip of the flange *(left)*.

• Reinstall the parts and wipe away any excess putty around the flange.

• If leaks persist, tighten all connections a quarter-turn.

FIX IT: Toilets

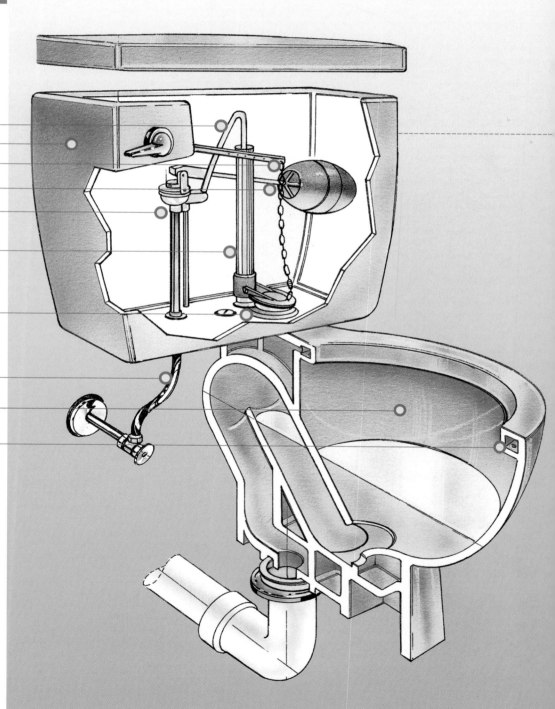

Refill Tube

Tank

Trip Lever and Lift Chain

Float Arm and Float Ball

Fill Valve (Ball Cock)

Overflow Pipe

Flush Valve and Seat

Supply Tube

Toilet Bowl

Flush Holes

Chapter 4

Contents

How They Work

Two mechanisms operate when a toilet is flushed: a flush valve and a fill valve (often called a ball cock). The illustration at right shows a flapper-style flush valve and a ball cock with a float ball. Two other common styles are shown on page 61.

Tripping the flush handle raises the flush valve, releasing water from the tank to the bowl. The rushing water creates a siphoning action in the bowl that forces waste water down the drain. As the tank empties, the falling water level lowers a float connected by a rod to the ball cock, opening the valve inside. When the flapper valve closes after the flush, the tank fills and the float, lifted by the rising water, closes the valve in the ball cock to keep the tank from overflowing.

Most flush toilets use conventional tank components and the parts needed for maintenance, repair, or replacement are readily available at hardware stores and home centers.

Troubleshooting

Problem	Solution
• **Toilet bowl overflows**	Use a plunger or auger **58** •
• **Toilet does not flush**	Tighten or replace the defective toilet handle **63** • Adjust or replace the lift chain **63** • Replace the flush mechanism **65** •
• **Toilet bowl drains sluggishly**	Remove the clog with a plunger or auger **58** • Raise the water level in the tank **62** •
• **Water runs continuously**	Lower water level in tank **62** • Adjust or replace lift chain or wire **63** • Repair or replace faulty ball cock **62** • Repair or replace worn flush valve **64** • Clean valve seat **63** • Replace valve seat assembly **64** •
• **Vibration when toilet tank fills**	Adjust the water level in the tank **62** • Adjust fill valve **62** • Replace fill valve **65** •
• **Water under the tank**	Tighten tank hold-down bolts **60** • Lower water level in tank **62** • Replace float valve **65** • Replace malfunctioning ball cock **65** • Insulate tank to prevent condensation **60** • Tighten nuts on supply tube (or hose), or replace leaking supply tube (or hose) **30** •
• **Seat loose**	Tighten loose bolts or replace the damaged seat **59** •

Before You Start

The toilet is the most heavily used plumbing fixture in the house, and while the porcelain bowl and tank will last indefinitely, the working parts inside will not.

CHECK THE WATER LEVEL

Although toilet parts eventually wear out, they are not the first place to look when confronted with a balky toilet. Flushing problems are often related to the water level in the toilet tank, and that is controlled by various fill mechanisms. The tank's water level should be 1/2 to 1 inch below the top of the overflow pipe. An incomplete flush may indicate a low water level, while water running over the top of the overflow pipe or leaking through the handle means that the water level is too high.

There are many things you can do to stop leaks when they occur and to restore a toilet's full flushing effectiveness. Sometimes, however, replacing the ball cock or flush valve offers the only lasting solution to flushing problems. In most cases, you can substitute parts made of plastic for the brass variety. Plastic is less expensive and, because this material does not corrode, plastic parts may last longer than brass ones.

Before You Start Tips:

⋯⋗ When working on a toilet, take care to place the tank cover in a safe place; porcelain cracks easily.

⋯⋗ Have plenty of rags, a sponge, and a pail nearby to catch water or clean up spills.

TOOLS

Pail

Plunger

Closet auger

Slip-joint pliers

Socket wrench and sockets

Hacksaw

Adjustable wrench

Locking pliers

MATERIALS

Rags

Sponge

Penetrating oil

Tape

Emery cloth

New toilet seat

Toilet tank insulation kit

Replacement flush parts or fill assembly

SAFETY FIRST

When bailing out a clogged toilet, wear rubber gloves to avoid infection.

Unclogging the Toilet

1. Using a plunger

A flange-type plunger *(inset)* fits into the toilet drain and exerts more pressure than the old-style cup-type plunger.

• If the bowl is full or overflowing, put on rubber gloves and use a plastic container to bail out half the water. If the bowl is empty, add water to half-full.

• Place the plunger over the drain opening (the larger one if there are two). Keeping the plunger below water level and firmly in place *(right)*, pump up and down rapidly 8 to 10 times. If the water rushes away, you may have removed the blockage.

• Use the plunger again to be sure the water is draining freely. Then pour in a pail of water and plunge one more time before flushing the toilet to refill the bowl.

• If the blockage remains after plunging, use a closet auger *(below)*.

Plunger

1

2. Using a closet auger

• Determine the direction in which to guide the auger. Some toilets are rear-draining *(right)*, while others are front-draining *(inset)*.

• Feed the curved tip of the auger into the drain opening. Crank clockwise until the auger tightens up, then crank in the other direction. When the auger tightens again, reverse the direction until the auger is as far in the drain as it will go.

• Pull the handle up and out to remove the auger. If it jams, push gently, then pull again. You may have to turn the handle as you pull up.

• Use a plunger *(Step 1)* to ensure that the drain runs freely.

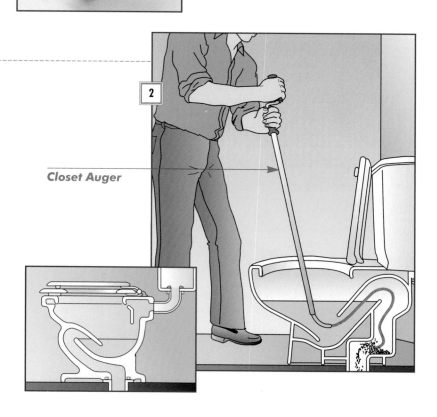

Closet Auger

2

Replacing the Toilet Seat

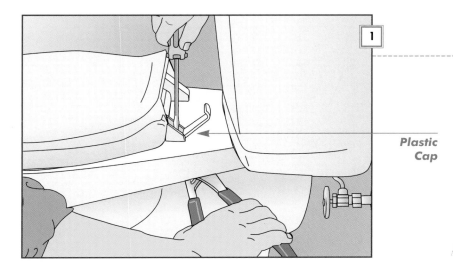

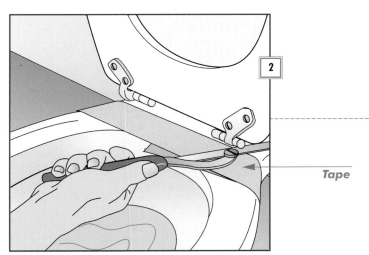

Plastic Cap

Tape

1. LOOSENING THE SEAT BOLTS

• The seat bolts may be hidden under plastic caps *(left)*. Pry open each cap (if any). Hold each bolt with a screwdriver, and unscrew the nut with a socket wrench or slip-joint pliers. (Some nuts have plastic "wings" that make pliers unnecessary.)

• To loosen corroded metal bolts, apply penetrating oil, wait overnight, then try again. If the bolts still will not budge, cut them off *(Step 2)*.

• Replace the seat and bolts, then hand tighten the nuts. Make sure the seat aligns with the bowl, then tighten the nuts one-quarter turn with a wrench or pliers.

2. CUTTING OFF THE SEAT BOLTS

• Place some tape or a thin piece of cardboard on the bowl to protect it from scratches.

• Cut through the bolts with a mini-hacksaw *(left)*.

SEATS AND SEAT MOUNTINGS

Toilet seats are available in a seemingly endless variety of colors, qualities, and materials. If comfort is your chief objective, choose a soft or padded seat. These seats do not last long, however, nor do seats with a composition-wood core coated with enamel. Solid plastic and solid wood seats tend to cost more initially, but they will last the longest.

Toilet seats come in two basic shapes and several different mounting arrangements. Take the old seat with you when you shop for a replacement, so you can verify that the distance between seat bolts is the same.

Insulating the Tank

1. LINING THE TANK WITH FOAM

Foam insulation inside a tank will often stop "sweating" caused by condensation. Insulation kits are available at building supply outlets and plumbing supply stores.

• Remove the tank cover and mark the water level inside the tank.

• Shut off the water supply, then flush the toilet, holding the handle to drain as much water as possible. Sponge up any remaining water in the tank, then finish drying the tank walls with a hair dryer.

• Using the glue provided with the kit, fasten the foam to the inside of the tank *(right)* so that the foam extends 1/2 inch above the water-level mark. Wait 24 hours before turning on the water to fill the tank.

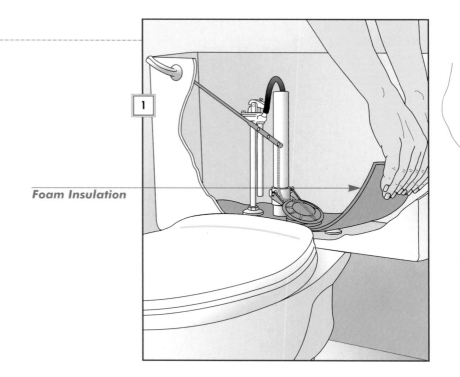

Foam Insulation

Stopping Leaks at the Tank

1. TIGHTENING THE HOLD-DOWN BOLTS

• With a wrench and screwdriver, tighten the tank hold-down bolts. (You needn't drain the tank.) Turn the nuts gradually; too much force may crack the tank.

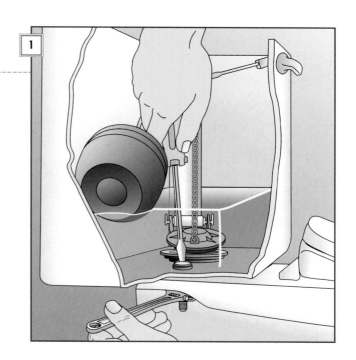

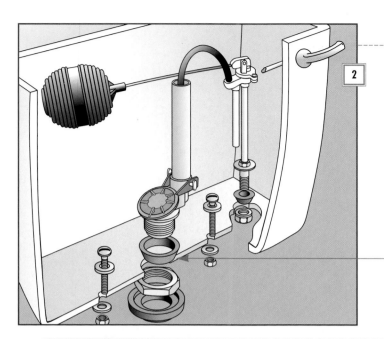

Gasket

2. FINDING THE SOURCE OF A LEAK

• If tightening the hold-down bolts has no effect, add a few drops of food coloring to the water in the tank. Wait, as much as a day, to see where the colored water leaks out.

• Drain the tank and replace any gaskets or washers at the leak, using the instructions for replacing a flush mechanism *(page 65)* and the illustration at left as a guide. Renewing the tank gasket requires dismounting the tank and unscrewing the lock nut below it.

TWO COMMON FLUSH AND FILL MECHANISMS

Upon removing the top of your toilet tank you may find, instead of the float-ball fill mechanism illustrated on page 54, one of the two types shown here. In the float-cup type *(right, top),* the flush handle lifts a flapper valve to begin the flush. As the flapper valve returns to its seat after the flush, the rising water floats a plastic cup that is connected by a shaft to a lever that turns off the water. In this model, the lever is near the bottom of the flush mechanism; other models have the lever near the top. Float-cup assemblies may not conform to the anti-siphon regulations of some local plumbing codes.

The tilt-cup flush valve *(right, bottom)* is essentially a time-delay mechanism. When the flush handle is pulled, it lifts the float cup and flapper valve to begin the flush. As the tank empties, water remains in the upturned cup but gradually drains through a small hole in its bottom. When the cup is empty, it falls forward, closing the flapper. The delay ensures that the tank will empty completely before the flapper closes.

FLOAT CUP

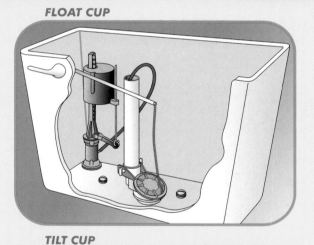

TILT CUP

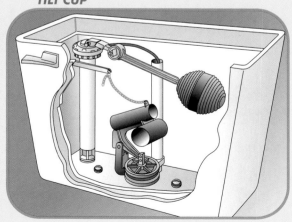

Adjusting the Water Level

BENDING A FLOAT ARM

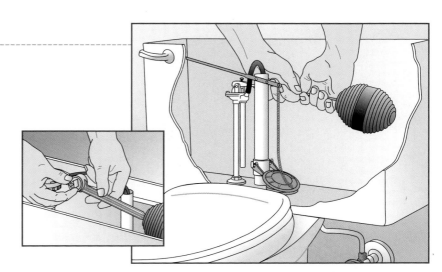

- To lower the water level, gently lift the float arm *(right)* and bend it down slightly to keep the water level 1/2 to 1 inch below the top of the overflow pipe.

- To raise the water level, bend the float arm to raise the float ball, making sure that the ball does not rub the tank.

- To raise or lower a plastic float arm *(inset)*, turn the knob at the ball cock.

RAISING OR LOWERING A FLOAT CUP

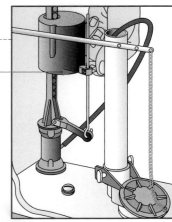

Retaining Clip

- Pinch the retaining clip *(right)* and slide the float cup 1/2 inch up or down with each adjustment.

- Lower the water level by lowering the float cup.

- Raise the water level by raising the cup.

REGULATING A METERED FILL VALVE

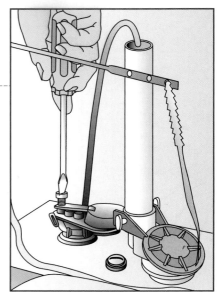

On this variety of flush valve, the water level is raised or lowered by turning the adjustment screw *(right)*.

- Lower the water level by turning the screw counterclockwise one half-turn at a time; raise the level by turning it clockwise.

- Test the toilet and readjust the water level if necessary.

Servicing the Flush Assembly

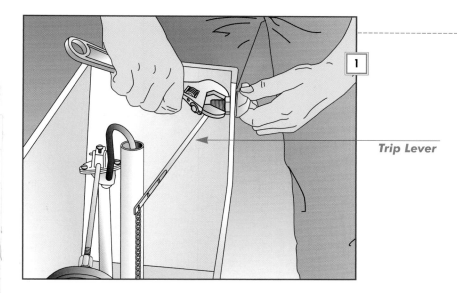

Trip Lever

1. ADJUSTING THE HANDLE

• Remove the tank cover. If the handle is loose, tighten the lock nut with a wrench *(right)*. If the nut won't budge, even after treatment with penetrating oil, cut through the handle shaft with a hacksaw and re-place the handle and trip lever.

• Unhook the chain from the trip lever and slide the trip lever, with the handle at-tached, through the hole in the tank.

• Scrub the handle threads with a tooth-brush and vinegar.

• Reinstall the assembly, then tighten the lock nut and adjust the lift chain *(below)*.

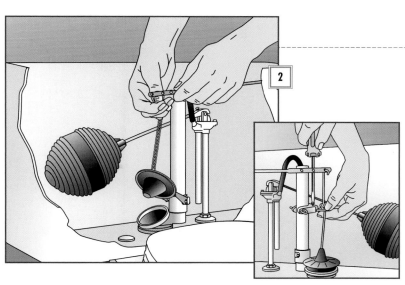

2. ADJUSTING THE LIFT CHAIN

• If the handle must be held down while the toilet is flushing, the chain may be too long. Shorten it by hooking it through a different hole in the trip lever *(right)*, or use long-nose pliers to open and remove some chain links.

• Some older flush assemblies have a lift wire *(inset)*, instead of a chain. If the wire binds against its guide, flushing is impaired. Loosen the guide with a screwdriver, then adjust it so that the flush valve falls freely onto the seat.

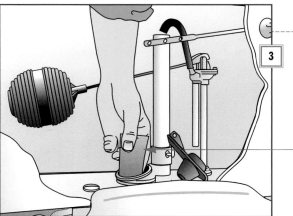

Emery Cloth

3. CLEANING THE VALVE SEAT

• Turn off the water supply and flush the toilet to drain the tank.

• Disconnect the refill tube and slide the flapper valve off the overflow pipe.

• With emery cloth, gently scour inside the seat and along its rim.

• Reassemble the mechanism, turn on the water supply, and flush to check for leaks.

4. REPLACING THE FLAPPER VALVE

• Unhook the refill tube and lift chain, then remove the flapper valve *(left)*.

• Buy a replacement that fits the valve seat in your tank and install it. If you cannot find a valve that fits your seat, or if leaks persist after replacing the flapper valve, replace the entire assembly *(box, below)*.

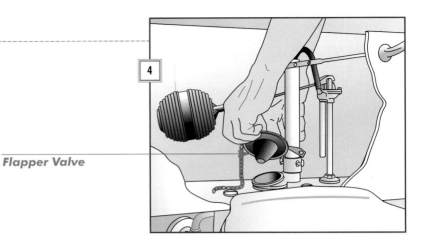

Flapper Valve

5. REPLACING THE FLOAT BALL

• Grasp the float arm with locking pliers and unscrew the float ball *(left)*. If it will not come off, use the pliers to unthread the float arm from the ball cock.

• Coating the threads with petroleum jelly, screw a new ball onto the float arm, then screw the arm back onto the ball cock.

• Flush the toilet and adjust the water level as necessary *(page 62)*.

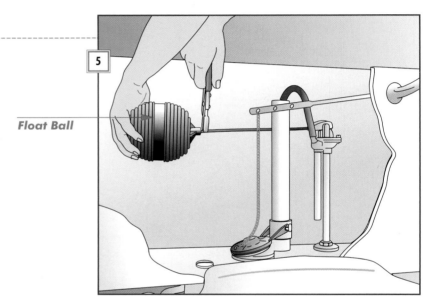

Float Ball

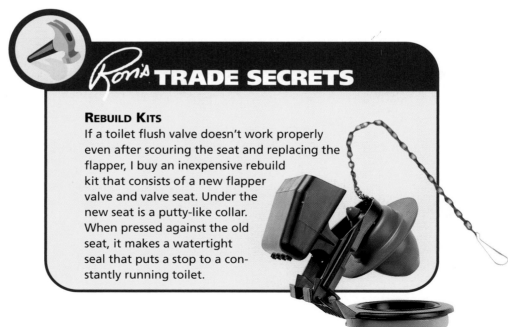

Ron's TRADE SECRETS

REBUILD KITS

If a toilet flush valve doesn't work properly even after scouring the seat and replacing the flapper, I buy an inexpensive rebuild kit that consists of a new flapper valve and valve seat. Under the new seat is a putty-like collar. When pressed against the old seat, it makes a watertight seal that puts a stop to a constantly running toilet.

Replacing the Flush Mechanism

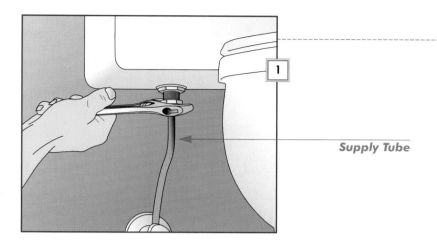

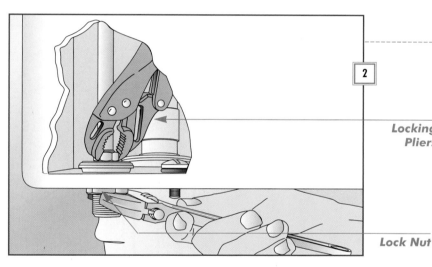

Supply Tube

Locking Pliers

Lock Nut

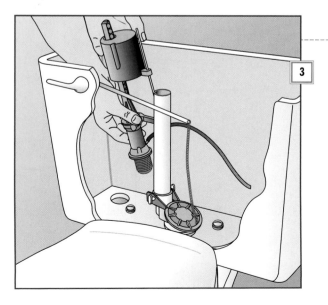

1. PREPARING THE TANK

• Shut off the water supply and flush the toilet.

• Remove the tank cover and place a container on the floor beneath the tank to catch water runoff. Sponge up any water remaining in the tank.

• With an adjustable wrench *(right)*, disconnect one end of the supply tube from the tank.

2. REMOVING THE FLUSH MECHANISM

• Attach locking pliers to the nut at the base of the ball cock *(right)*, then loosen the lock nut under the ball cock with an adjustable wrench. Lift the ball cock up from the tank.

• Use a bristle brush or a nylon scrubbing pad to clean the opening in the tank where the ball cock was.

3. INSTALLING NEW PARTS

• Place a rope of plumber's putty around the cone-shaped washer of the new ball cock, then set the assembly into the tank *(right)* and tighten the lock nut snugly.

• Place the refill tube into the overflow pipe, reconnect the supply tube, and slowly turn on the water. Slightly tighten the lock nut if it leaks.

• Adjust the water level in the tank as necessary *(page 62)*.

FIX IT: Tubs and Showers

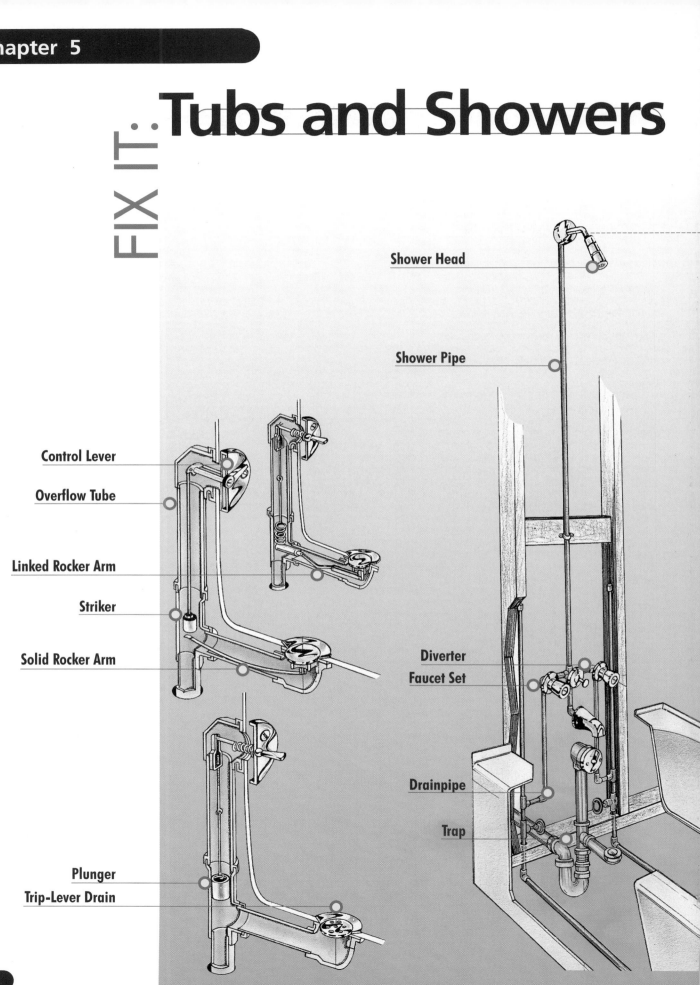

Shower Head

Shower Pipe

Control Lever

Overflow Tube

Linked Rocker Arm

Striker

Solid Rocker Arm

Diverter

Faucet Set

Drainpipe

Trap

Plunger

Trip-Lever Drain

Chapter 5

How They Work

Concealed behind a wall of tile or a skin of fiberglass is a system of supply pipes and drains designed for long-term durability. The heart of the supply system is the faucet set, fed by separate hot-and cold-water pipes. Single- or double-handle faucet controls determine the temperature and volume of water that is blended in the faucet body. A diverter acts like a switch to direct the flow either through the tub spout or up through piping to the shower head. The controls will eventually succumb to wear, while the diverter and shower head are more prone to problems caused by minerals and other contaminants in the water supply.

Drains easily clog, but are mechanically simple and easy to clean. Pop-up drains *(far left, top)* feature a stopper that settles over the drain opening to block water, or rises slightly to release it. Because many of its parts are directly in the path of draining water, they accumulate debris quickly and require frequent cleaning. A trip-lever drain *(far left, bottom)* is capped by a strainer plate instead of a stopper. Inside the overflow tube, a hollow metal plunger moves up or down to block or unblock the end of the drain.

Contents

Troubleshooting

Problem	Solution
• **Water seeps from bathtub (pop-up drains)**	Remove, clean, and adjust stopper and lift assembly **70** • Adjust striker or replace lift assembly **71** •
• **Control lever won't stay in position**	Adjust striker **71** •
• **Water drains too slowly from bathtub, or is blocked entirely**	Clean stopper **70** • Clean and adjust lift assembly **71** • Open drainpipe with plunger, auger, or hose **72** •
• **Water incompletely diverted from spout to shower**	Replace tub spout diverter **87, 88** • Clean diverter **87, 88** • Replace O-rings **87** •
• **Water leaks around diverter handle (push-pull diverters)**	Replace O-rings or diverter **87** •
• **Shower head leaks**	Tighten shower head to arm or seal joint with pipe tape **89** •
• **Drainage system is sluggish; household drains smell bad**	Have septic system checked **91** •
• **Faucet handle leaks**	Replace worn O-rings or stem on two-handle faucets **78** • Tighten adjusting ring on lever faucets **78** •
• **Water leaks from tub spout or shower head (double-handle faucets)**	Rebuild a two-handle faucet **74** • Replace worn parts on ball faucet **77** • Replace cartridge on cartridge faucet **79** • Replace worn faucet set **80, 82** •

Before You Start

About one-third of the water used by a typical household pours down the bathtub drain. Little wonder, then, that drains clog from time to time and fixtures may drip.

You can do much to prevent or delay the formation of clogs (*page 73*). However, once a drain has become blocked, it is best opened with a plunger or an auger; chemical drain openers contain caustic compounds that can weaken the drainpipe. If you use a drain cleaner, check the label for cautions, and wear rubber gloves and goggles.

Fixing leaky faucets in the tub and shower parallels similar repairs in the kitchen and bathroom. The main difference is one of access. Most tub-and-shower plumbing is inside a wall or under the floor. If the wall has no access panel behind the faucets, you may well simplify a repair by creating one (*page 84*). Likewise, reaching a clogged trap when a plunger or auger has no effect often calls for knocking a hole in the ceiling of the room below, a job most people leave to professionals.

IMPORTANT FIRST STEPS

Before repairing supply pipes and fixtures, turn off the water supply. Shutoff valves can be found behind an access panel, or directly below the tub in the basement. If you find none, close the main shutoff valve for the house.

When working on chromed or brass parts, protect them from the bite of a wrench by wrapping them with tape. Also, be sure to have a pail and some rags close at hand to soak up any water that spills from the system.

TOOLS

Pliers

Plunger

Auger

Hose

Expansion nozzle

Screwdriver

Hammer

Cold chisel

Socket wrench and deep socket

Adjustable wrench

Valve-seat wrench or large hex wrench

Minihacksaw

Propane torch

Tube cutter

Long-nose pliers

MATERIALS

Fine steel wool

Rags

Rubber ball

Pail

O-rings

Stem washers

Valve-stem grease

Valve seat

Pipe-joint tape

Faucet repair kit

Solder

Flux

Before You StartTips:

⋯⋗ Cover the drain with a bath mat to protect the tub from dropped tools and other hazards, and to prevent the loss of small parts.

⋯⋗ To loosen or remove stubborn threaded-steel parts, spray on penetrating oil and wait 15 minutes before trying again.

⋯⋗ Remove corrosion with vinegar and steel wool or by scrubbing with an old toothbrush.

⋯⋗ When disassembling a faucet or shower head, stick the many small parts to a strip of masking tape in the same order in which you remove them.

⋯⋗ When replacing an old shower head with a new one, consider upgrading it to one that saves water or has an adjustable spray.

Fixing the Drain Assembly

1. CLEANING THE POP-UP STOPPER

• To remove the stopper, turn the control lever to open the drain, pull up the stopper *(right)*, and work the rocker arm clear of the drain opening.

• Remove accumulated hair and soap, then clean the assembly with fine steel wool.

• Feed the rocker arm back into the drain with the arm curving downward in the drainpipe. Wiggle a linked rocker arm (the arm is hinged, not rigid) back and forth until it sits back in place.

• If drain problems persist, go to Step 2.

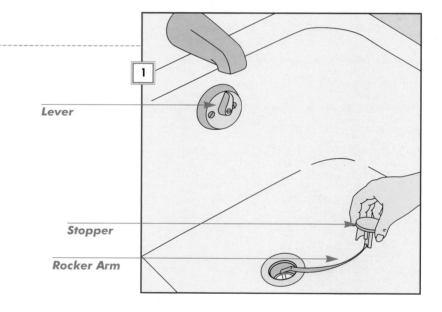

Lever

Stopper

Rocker Arm

2. REMOVE THE LIFT ASSEMBLY

• Cover the drain with a bath mat to protect the tub and prevent loss of parts.

• Remove the screws securing the overflow plate to the tub, then pull the lift assembly up through the overflow opening *(right)*.

• Wash debris from the assembly. Scrub away corrosion using vinegar and either steel wool or an old toothbrush.

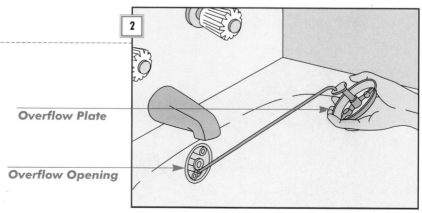

Overflow Plate

Overflow Opening

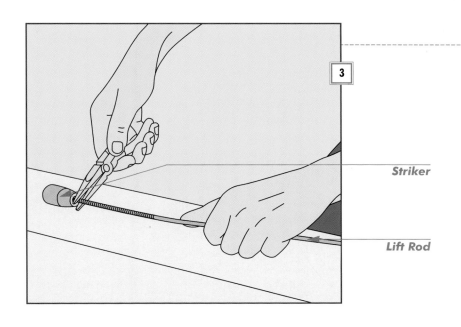

3. REPOSITIONING THE STRIKER

• Remove the lift rod from the overflow plate, if possible. Loosen the lock nut that holds the striker in place *(right)*.

• Rotate the striker down the threaded rod to lengthen the lift assembly (and raise the stopper higher), or up to shorten it. Then, retighten the lock nut, reassemble, and test.

• If problems persist, replace the drain assembly. When a drain assembly of the same make and model as the old one is not available, substitute a rubber plug, or a new flange and stopper *(below)*.

Striker

Lift Rod

Replacing a Drain Flange

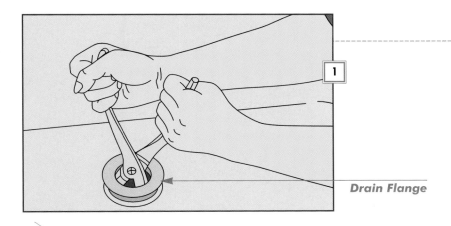

1. REMOVING THE OLD FLANGE

• Pull out the pop-up stopper (or unscrew the trip-lever strainer), then remove the overflow plate and lift assembly. Disconnect or cut off the lift rod. Screw the overflow plate back in place.

• With the stopper or strainer removed, use pliers to unscrew the drain flange *(left)*.

Drain Flange

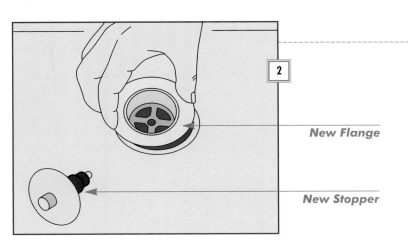

2. INSTALLING THE NEW FLANGE

• Take the flange to a plumbing supply store for a replacement whose threads match those of the original. It will come with a matching stopper.

• Apply a strip of plumber's putty under the lip of the new flange.

• Screw the flange into the drain opening and thread the metal stopper into the crosspiece. Depress the stopper once to close the drain, again to open it.

New Flange

New Stopper

Clearing a Clogged Tub

A PLUNGER

• If there is a tub stopper, remove it and the overflow plate *(page 70)*. Plug the over-flow opening with a large, wet rag.

• Run enough water into the tub to cover the plunger cup.

• Work the plunger vigorously up and down over the drain opening *(right)*. Continue plunging for several minutes.

• If the clog remains, clear the drain with one of the following methods.

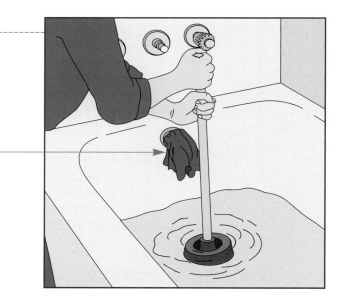

Wet Rag

AN AUGER

• Have a pail ready to catch any debris snagged by the end of the auger.

• In a shower stall, pry up or unscrew the strainer and work through the drain open-ing. In a bathtub *(right)*, remove the stop-per and the lift assembly *(page 70)*, and feed the auger down the overflow tube.

• Maneuver the auger around the corners in the drain, rotating it clockwise to break up the clog. Slowly remove the auger, then run water to test the drain.

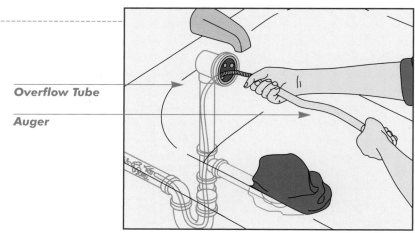

Overflow Tube

Auger

A HOSE

• If a hose won't reach the tub through a window, attach it to an indoor faucet using a threaded adapter.

• Close other drains in the room, feed the hose down the overflow tube and pack rags tightly around it *(right)*. Press firmly on a plug or a rubber ball to seal the drain.

• Hold the hose firmly while a helper turns the water on full force and then off again several times to flush the blockage.

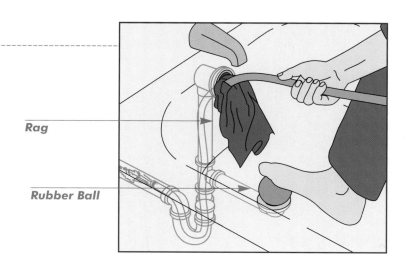

Rag

Rubber Ball

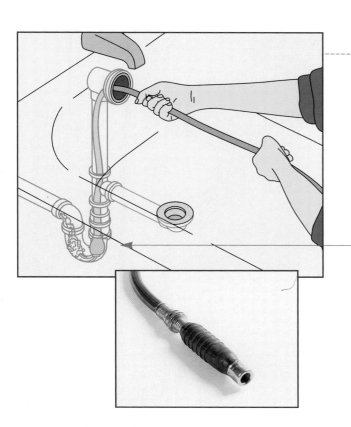

AN EXPANSION NOZZLE

• Measure the drain opening to determine which expansion nozzle will fit. Attach the nozzle *(inset)* to the hose and the hose to a faucet.

• Seal off all nearby drains. Insert the nozzle *(left)*, then turn the water slowly to full force, inflating the nozzle to seal the drain.

• After 10 seconds or so, turn the water off. Detach the hose from the faucet to let the nozzle deflate before removing it.

Nozzle

COPING WITH CLOGS

The most common cause of clogged drains in bathtubs and showers is hair and soap sludge. The best way to handle such clogs is to keep them from developing in the first place. One way is to replace the pop-up *(page 70)* with a rubber stopper and basket strainer *(right)*. Regularly remove and clean the strainer.

If a drain begins to empty slowly, act immediately to clear the developing clog. Examine the lift assembly and the underside of the stopper for an accumulation of pup-up as the cause of the developing stoppage. If the problem persists, clear the drain with one of the methods shown here.

Rebuilding Two-Handle Faucets

1. PRYING OFF THE HANDLE COVER

• First, turn off the water supply and open the faucets.

• Close the drain and cover it with a bath mat to protect the tub and to prevent loss of small parts.

• Pry off the handle cover using a screw-driver *(right)* or a knife.

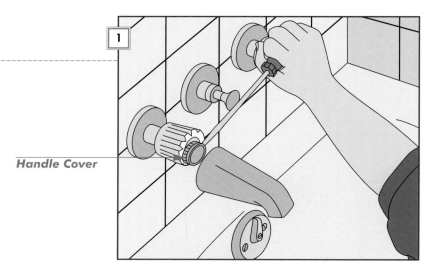

Handle Cover

2. REMOVING THE HANDLES

• Remove the retaining screw, then pull off the handle *(right)*.

• If the handle is stubborn, pour hot water over it and carefully pry up the base with a screwdriver. Alternatively, use a faucet-handle puller, available at hardware stores.

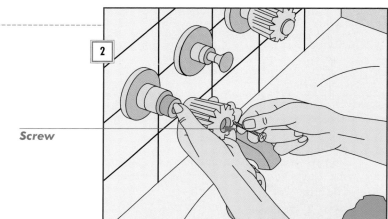

Screw

3. REMOVING THE ESCUTCHEON

• If there is a setscrew in the escutcheon, loosen it with a hex wrench. Use a screw-driver to pry the escutcheon from the wall *(right)*. Be careful not to damage ceramic tiles on the wall.

• Tap a stubborn escutcheon with the screwdriver or spray it with penetrating oil and wait 15 minutes before trying again.

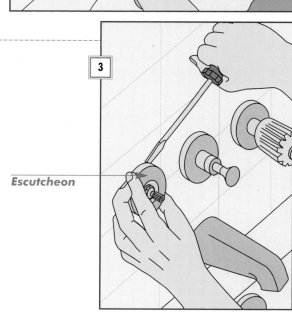

Escutcheon

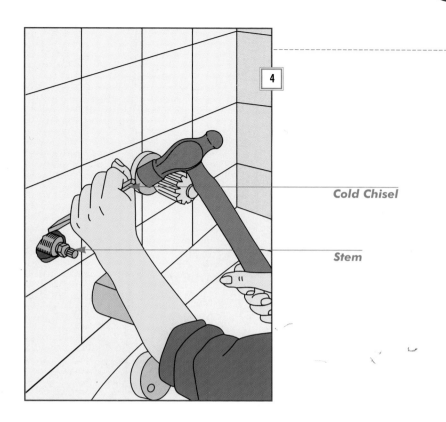

4. TRIMMING BACK THE WALL SURFACE

• If the bonnet nut on the stem is below the surface of the wall, trim back the tile, if any, and use a cold chisel and hammer to chip out any plaster or concrete.

Cold Chisel

Stem

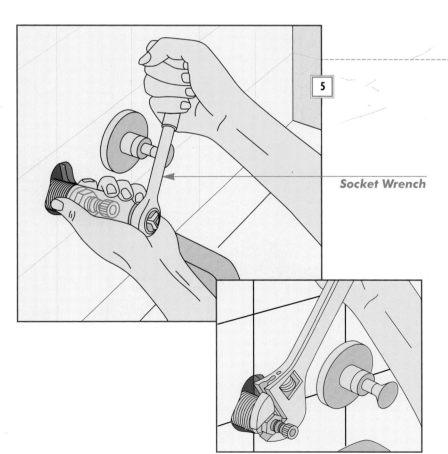

Socket Wrench

5. REMOVING THE STEM

Use a socket wrench with a deep socket to remove the bonnet nut and stem *(left)*.

• Fit the socket over the stem and onto the nut, then turn the wrench counterclockwise. If the stem stays in the faucet, which it may in old sets, unscrew and remove it. If necessary, spray on penetrating oil and wait 15 minutes before trying again.

• If you are using an adjustable wrench *(inset)*, be careful not to damage the soft brass bonnet nut.

6. CURING LEAKS AROUND THE HANDLE

• Separate the stem from the nut *(right)*. For better leverage, reattach the faucet handle.

• Pry off and replace the O-rings. This will probably cure handle leaks, but as long as the stem is off you might as well replace other serviceable parts *(following steps)*. If the stem appears worn, pitted, or corroded, replace the entire assembly.

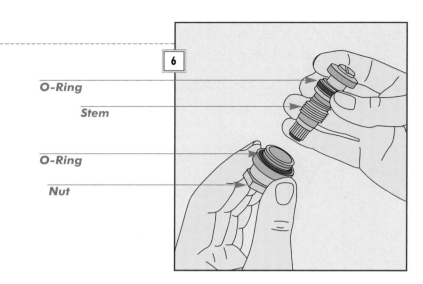

O-Ring
Stem
O-Ring
Nut

7. CURING LEAKS FROM THE TUB SPOUT OR SHOWER HEAD

• Remove the screw that holds the seat washer to the stem *(right)*. Replace the old washer with a new one of the same size.

• While the stem is out of the faucet body, inspect the valve seat in the wall. If it appears damaged, replace it *(next step)*. Lubricate the new stem with valve-stem grease before installing it.

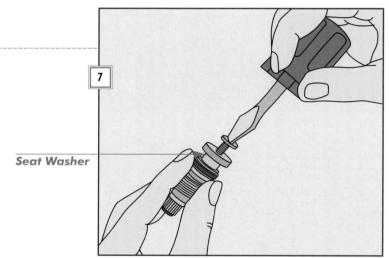

Seat Washer

8. REPLACING THE SEAT

Some seats are built into the faucet and cannot be removed for replacement. Instead, its surface must be smoothed with a tool called a valve-seat dresser, available at plumbing supply stores. To replace a removable valve seat:

• Unscrew the worn seat with a valve-seat wrench or a large hex wrench *(right)*, depending on the shape of the hole in the center of the seat.

• Buy an exact replacement, lubricate the threads with pipe-joint compound, then screw it into the faucet body.

• Reinstall all parts and test.

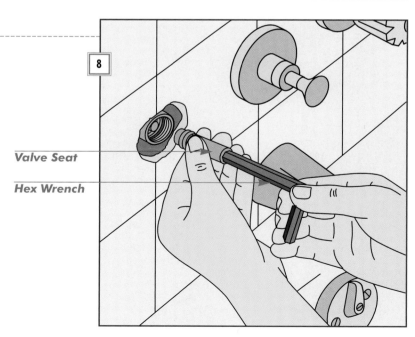

Valve Seat
Hex Wrench

Repairing Ball Faucets

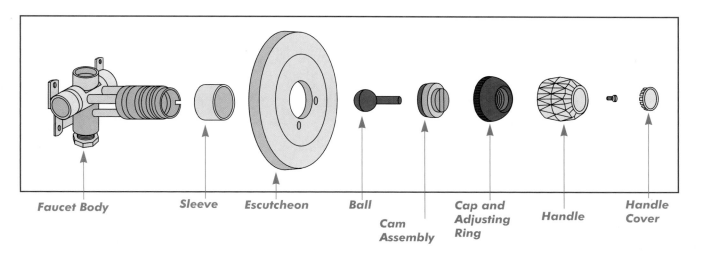

Faucet Body   **Sleeve**   **Escutcheon**   **Ball**   **Cam Assembly**   **Cap and Adjusting Ring**   **Handle**   **Handle Cover**

ANATOMY

A faucet such as the one above works much like a rotating-ball sink faucet *(page 20)*. Valve seats and springs, located inside the faucet body and available as a repair kit, are the prime suspects when the spout or shower head leaks.

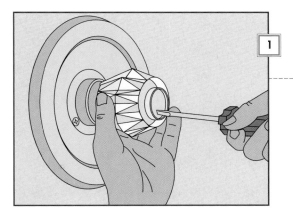

1. REMOVING THE HANDLE

• If a single-lever ball-type faucet leaks, turn off the water supply and open the faucet.

• Pry off the handle cover with knife or narrow-blade screwdriver *(left)*. Remove the handle screw, and lift off the handle.

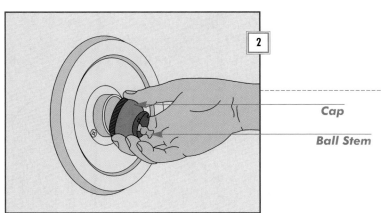

Cap

Ball Stem

2. REMOVING THE CAP AND BALL

• Unscrew and lift off the cap *(left)*.

• Pull the ball stem to remove the cam-and-ball assembly that controls the mixture of hot and cold water.

3. REPLACING THE SEATS AND SPRINGS

• Use a small screwdriver or a nail to extract the rubber seats and springs from the two small sockets in the faucet body *(right)*.

• Set in new springs. If the springs are cone-shaped, insert the large end first.

• Place new seats over the springs.

• To reassemble the faucet, set the ball in the faucet body, adjust the cam assembly over the ball, and screw on the cap.

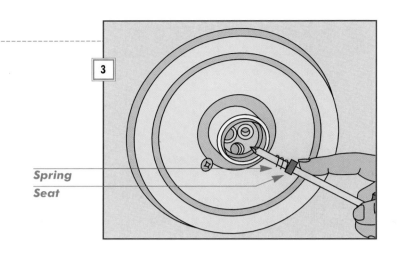

3

Spring

Seat

4. TIGHTENING THE ADJUSTING RING

• Using pliers (or a special wrench often included in the ball-type repair kit), turn the adjusting ring inside the cap clockwise *(right)*.

• Turn on the water supply, then use the ball stem to turn on the faucet. If water leaks from around the stem, tighten the adjusting ring.

• Reinstall the handle once the leaks have been corrected.

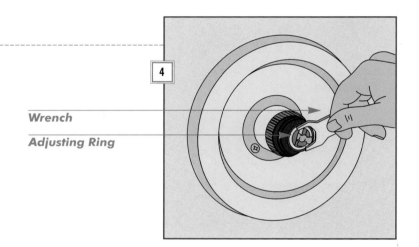

4

Wrench

Adjusting Ring

TEMPERATURE BALANCING FAUCETS

Hot water can pose a danger, especially to the very young and very old. That's why many modern shower faucets are equipped with devices that keep the water at a safe temperature. These temperature-balancing or "anti-scald" faucets are sometimes required in new houses, but they make equal sense in old ones.

Unfortunately, you can't always tell the difference between anti-scald faucets and others; what makes them work is hidden.

Some models work by preventing the hot water handle from being turned fully ON others prevent the cold water valve from closing completely. More accurate are those that contain a bimetal spring that responds to a preset temperature, closing or opening the hot-water valves as needed. The perforated valve at right is from an antiscald faucet; the other is from a standard faucet.

Fixing Cartridge Faucets

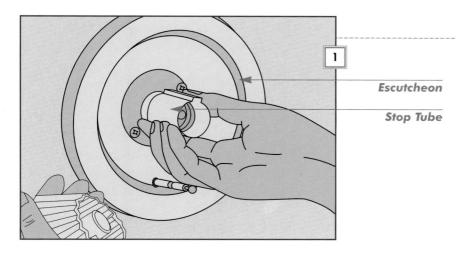

Escutcheon

Stop Tube

1. GETTING AT THE CARTRIDGE STEM

• If a single lever cartridge-type faucet leaks, turn off the water supply and open the faucet to empty the supply lines. Use a screwdriver or knife to pry off the handle cover, exposing the handle screw; remove the screw.

• Lift off the handle and pull off the stop tube *(left)* to reveal the stem of the cartridge and its retainer clip, which secures the cartridge to the faucet body. (On some faucets, the escutcheon must be lifted off first; look for retaining screws or a setscrew.)

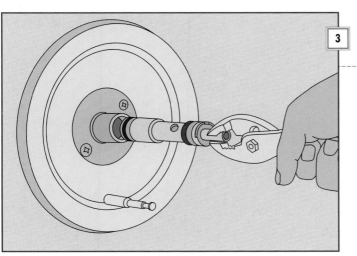

Retainer Clip

2. FREEING THE CARTRIDGE

• Using a small screwdriver or a nail, carefully pry up the cartridge retainer clip *(left)*. To prevent the clip from springing loose, hold a finger against it as it lifts free of the cartridge.

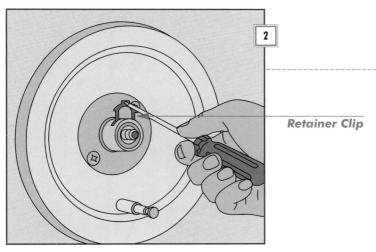

3. REPLACING THE CARTRIDGE

• Grip the cartridge stem with pliers *(left),* then pull it from the faucet body by rotating it from side to side.

• Insert the new cartridge with the flat part of the stem facing up (otherwise, the hot and cold water will be reversed).

• Reinstall the retainer clip, the stop tube, and the handle.

Replacing a Lever Faucet

1. GAINING ACCESS TO THE FAUCET BODY

● Turn off the water supply and open the faucet. Remove the handle *(page 79)*, then unscrew the escutcheon and lift it off *(right)*.

● If there is an access panel behind the faucet, remove the panel and replace the faucet from the back, adapting the procedure for a double-handle faucet (shown on pages 82-83). Otherwise, it may be possible to replace the faucet set from the tub side, as shown in the following steps.

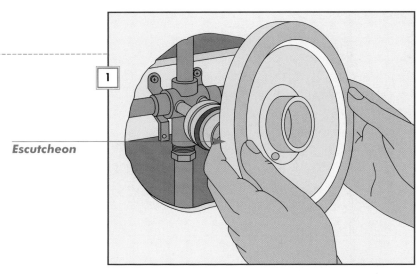

Escutcheon

2. REMOVING THE FAUCET SET

● Cut the copper shower pipe, spout pipe, and supply pipes with a minihacksaw *(right)*. If necessary, chip away tiles or wallboard for better access, but be careful not to remove more than the escutcheon will cover.

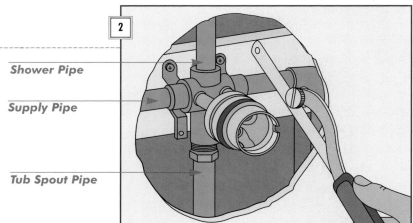

Shower Pipe

Supply Pipe

Tub Spout Pipe

3. UNSCREWING THE FAUCET BODY

● Remove the screws that secure the faucet body to the wooden crosspiece behind it *(right)*. Lift the faucet set out of the wall.

● Purchase a replacement faucet of the same make and model as the old one, or at least the same size. Also, buy a piece of copper pipe 2 inches longer than the combined length of the pipe stubs in the old faucet plus four slip couplings.

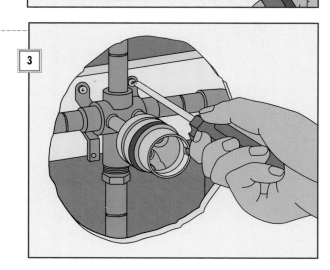

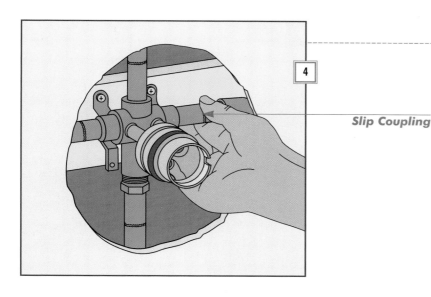

4

Slip Coupling

4. RECONNECTING THE PIPES

● Cut the new pipe into four lengths to connect the shower pipe, tub spout pipe, and supply pipes, and ream them carefully. (To cut, connect, and solder copper pipe— or to replace copper with plastic pipe— see Chapter 6.)

● Position the faucet in the wall, using the slip couplings to bridge the joints between pipes *(left)*.

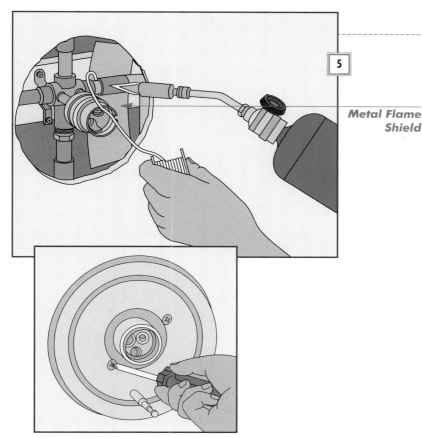

5

Metal Flame Shield

5. COMPLETING THE INSTALLATION

● Secure the faucet body by screwing it to the wood crosspiece.

● Remove all plastic and rubber parts from the faucet body to prevent them from heat damage, then solder the joints at the couplings *(left)* and at the faucet body.

● Screw on the escutcheon *(inset)* and reconnect the handle. Turn on the water supply and run water through the faucet. If the pipes leak around the joints, desolder the joints *(page 100)* and resolder them. After using the faucet for several days, remove the handle and escutcheon to check again for leaks.

CAUTION: Whenever you solder pipes in the wall, place a flame shield over the studs or other flammable material. Flame shields are available at hardware stores and plumbing supply stores. Keep a fire extinguisher nearby as well.

Replacing a Two-Handle Faucet

1. EXPOSING THE FAUCET BODY

• If there is an access panel on the other side of the wall, remove it *(right)*. If there is no panel, install one *(page 84)*.

• Turn off the water supply and open the faucets. Remove all handles and escutcheons, and the tub spout.

Faucet Body

Panel

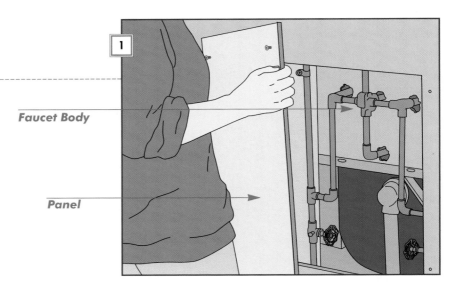

2. CUTTING THE PIPES

• Cut the shower pipe, spout pipe, and supply pipes with a standard tube cutter where they can be easily connected to the new faucet body with slip couplings and elbows *(right)*. Use a small tube cutter *(inset)* or a minihacksaw for cramped locations.

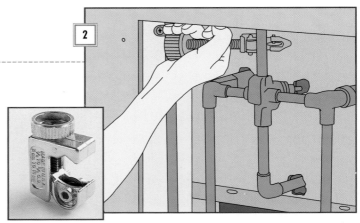

3. REMOVING THE OLD FAUCET BODY

• Ease the faucet body away from the wall *(right)*. If it is stuck, chip away at the grout and plaster from the front or pull off the silicone caulking with long-nose pliers.

• Take the faucet body, with the cut pipes still in it, to a plumbing supply store. Buy a replacement faucet body of the same make and model as the old one, or at least the same size.

• To calculate the amount of new pipe required, add the lengths exposed in the old faucet plus half an inch for each. Buy copper pipe and the necessary elbows and couplings.

Coupling

4. CONNECTING THE SHOWER PIPE

• Cut the lengths of pipe needed to connect the new faucet to the shower pipe and supply pipes.

• Solder the pipes into the new faucet. (You may need a helper to hold them in place while you work.)

• Position the faucet body against the back of the wall *(left)*, fitting the faucets and diverter (if there is one) through the holes in the wall.

• Use a coupling at the joint between the shower pipe and the faucet body.

5. CONNECTING THE SUPPLY LINES

• Holding the faucet in place, connect one supply pipe to the faucet body *(left)*, using a slip coupling or an elbow, depending on the configuration of the pipes.

• Connect the other supply pipe.

6. FINISHING THE JOB

• Solder each of the assembled joints *(pages 101-103)*.

• To prevent seepage behind the wall, seal around the pipes with plumber's putty or silicone.

• Install the escutcheons, attach the faucet handles and diverter knob, if any, and re-connect the tub spout.

• Leave the access panel open for a few days to check for leaks. If the pipes leak around the joints, desolder the joints *(page 100)* and resolder them.

Adding an Access Panel

1. MAKING THE CUTOUT

- To determine the location of the faucet set within the wall, measure from the floor up to the faucet, then from the corner of the tub wall out to the faucet. Transfer these measurements to the wall behind the bathtub.

- Use a hammer and an old chisel to make a hole at the appropriate spot in the wall.

- With a keyhole saw or drywall saw, cut from the hole out to the stud, then down *(right)*. Use a straightedge to draw a rectangular box, stud to stud and about two feet high, then continue cutting. Be careful not to cut any pipes or wiring.

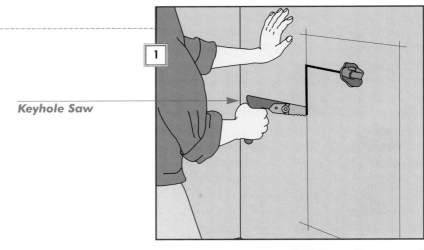

Keyhole Saw

2. ADDING CLEATS

- Nail a wooden cleat to the studs at each corner of the opening to accept screws for attaching a plywood panel.

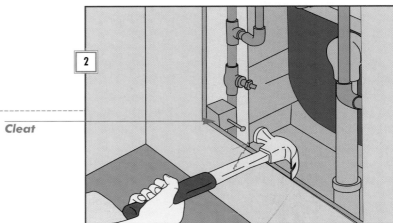

Cleat

3. SECURING THE NEW PANEL

- Cut a plywood panel slightly larger than the hole in the wall.

- With screws, attach the access panel to the cleats *(right)*.

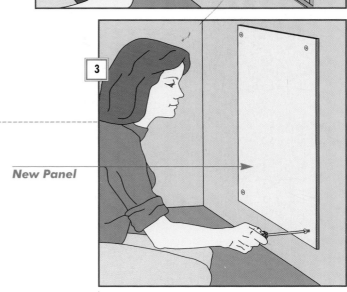

New Panel

Replacing a Slip-Fit Tub Spout

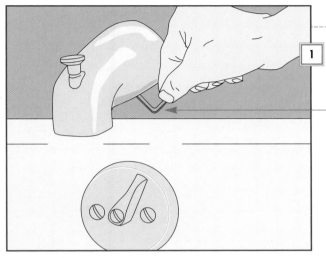

1. REMOVING THE OLD SPOUT

- Using a hex wrench, loosen the clamp screw on the underside of the spout *(left)*.

- Grasp the spout firmly and twist it back and forth off the pipe.

Hex Wrench

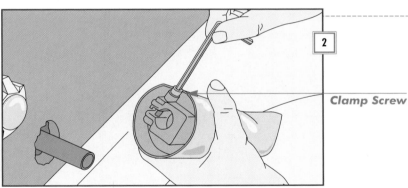

2. MOUNTING THE NEW SPOUT

- Loosen the clamp screw on the new spout with a hex wrench *(left)*, then twist the spout onto the copper pipe.

- Turn the spout so that the clamp screw faces up, and partially tighten the screw. Twist the spout into position and finish tightening the screw with the hex wrench.

Clamp Screw

Ron's TRADE SECRETS

TUB SPOUTS AND OLD HOUSES

Old houses are full of surprises, and I've found that goes double for plumbing repairs. Replacing an old tub spout can be a nightmare if the spout is too short. Now, though, I never worry. If the spout pipe is a little too short, I head for a home center or hardware store for a spout retrofit kit. Along with the new spout, the kit contains screw-together fittings that adapt to just about any spout pipe, no matter what length, diameter, or material.

Replacing a Screw-On Tub Spout

1. LOOSENING THE OLD SPOUT

● Check under the spout for a setscrew; if there is one, it is a slip-fit spout *(page 85)*. Otherwise, after protecting the spout with a rag, grip the spout with a pipe wrench *(right)*, and turn counterclockwise. If the spout will not move and there is access from behind, apply penetrating oil, wait 15 minutes, and try again.

● Do not use too much force or you might damage the plumbing behind the wall.

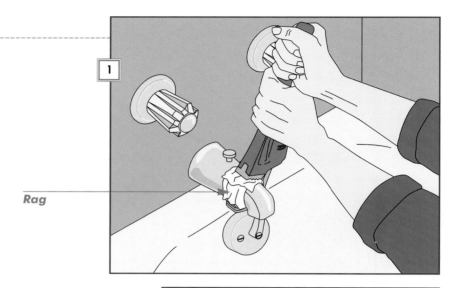

Rag

2. REMOVING THE SPOUT

● Twist the loosened spout off the nipple by hand *(right)* or off the adapter.

● Buy a compatible replacement spout.

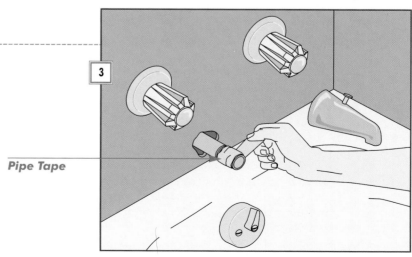

Pipe Nipple

3. MOUNTING THE NEW TUB SPOUT

● Clean the threads of the nipple with a wire brush. Apply pipe tape to the threads *(right)*, unless the directions on the new spout advise otherwise. Spread a bead of silicone sealant on the base of the spout.

● Thread the new tub spout onto the nipple and tighten it by hand, then wrap the new tub spout with a rag and tighten it with a wrench.

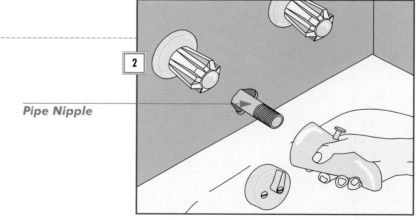

Pipe Tape

Push-Pull Diverters (Two-handle faucets)

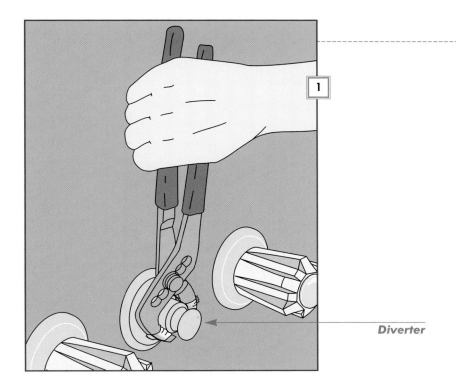

Diverter

1. LOOSENING THE DIVERTER

● Turn off the water supply and open the faucets. Close the drain and set a bath mat down to prevent loss of small parts down the drain and to protect the tub.

● Use pliers, jaws wrapped liberally with tape, to unscrew the diverter and expose two O-rings.

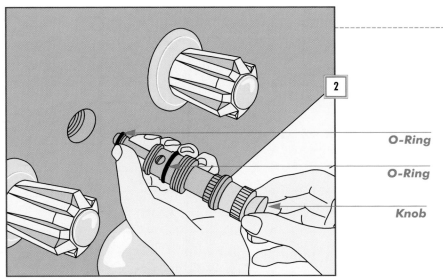

O-Ring

O-Ring

Knob

2. CLEANING THE MECHANISM

● Unscrew the diverter knob by hand *(left)*, to reveal another O-ring and a spring inside the housing.

● Clean the spring and lubricate it lightly with valve-stem grease.

● Replace all of the O-rings if any of them appear cracked or worn.

● Reinstall the diverter knob, then screw the entire assembly back into place. Turn on the water supply.

● If problems persist, replace the entire diverter with an identical part.

Push-Pull Diverters (Lever faucets)

1. REMOVING THE DIVERTER

• Remove the faucet handle and escutcheon *(right)*. Unscrew the diverter using an adjustable wrench or channel-joint pliers, depending on the type of diverter.

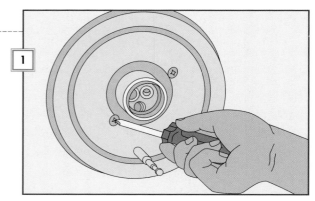

2. CLEANING THE MECHANISM

• If water is not being properly diverted from the tub spout to the shower head, clean any sediment off the diverter with vinegar and an old toothbrush *(right)*.

• When water leaks around the diverter, or if its parts are worn, replace the entire mechanism with one of the same make.

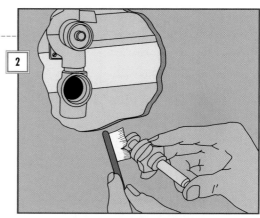

Ron's TRADE SECRETS

EMERGENCY DIVERTER REPAIRS

Like many plumbing parts, diverters rarely pick a convenient time to give up. In some cases, the base of the diverter knob corrodes so badly that pressure build-up behind it will send it shooting out into the tub, just as you're stepping into the shower. Believe me, I know. If you can't stand the thought of a bath and don't have time for a permanent repair, reach for the duct tape. Turn off the water and wipe off the area around the diverter, then slip the knob (and what's left of the diverter stem and spring) back into place. Pull the knob out to the shower position, secure it in place with a few turns of duct tape *(right),* and turn the water back on. It's not an elegant repair, but it'll get you to work on time.

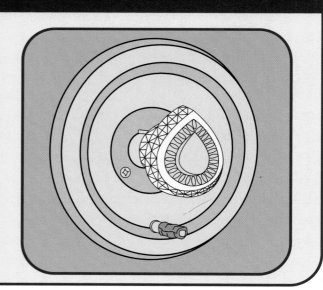

Repairing a Shower Head

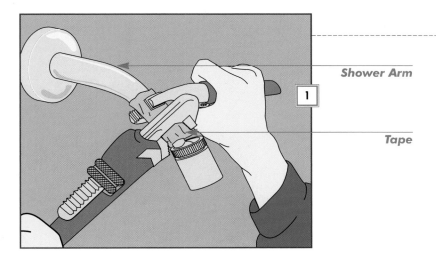

Shower Arm

1

Tape

1. REMOVING THE SHOWER HEAD

● Wrap the shower arm and shower head collar in masking tape and turn the collar counterclockwise with a pipe wrench (or use a strap wrench or rubber-jawed pliers, which require no tape). For greater lever-age, grip the shower arm with one wrench and turn the collar counterclockwise with another wrench *(left)*.

● Finish unscrewing the loosened shower head by hand. Don't let it drop into the tub because it will chip the surface.

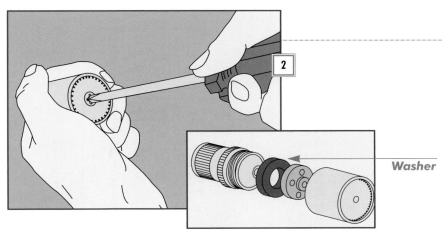

2

Washer

2. DISASSEMBLING THE UNIT

● Remove the screw or the knob that secures the faceplate to the shower head *(left)*.

● Unscrew the collar from the shower head to reveal the swivel ball, and pry out the washer *(inset)*.

● Clean the parts *(Step 3)*, or replace the shower head if the parts appear badly worn or corroded *(Step 4)*.

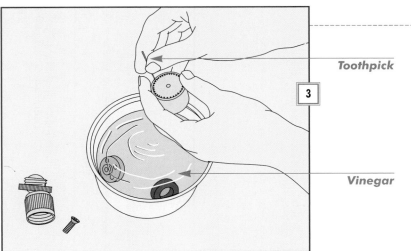

Toothpick

3

Vinegar

3. CLEANING THE PIECES

● Soak the shower head parts overnight in vinegar. Scrub with steel wool and an old toothbrush, and clear the spray holes with a needle or toothpick *(left)*.

● If possible, buy replacements for the worn parts of an expensive shower head rather than buy a new one. Lubricate the swivel ball with petroleum jelly or silicone lubricant. To reassemble the shower head, reverse the order of disassembly in Step 2.

4. INSTALLING THE SHOWER HEAD

• With a wire brush, clean the pipe threads at the end of the shower arm.

• Apply pipe tape to the threads to seal the joint *(right)*.

• Hand-tighten the shower head on the shower arm.

• Turn on the water to test the shower; if it leaks, tighten the head an additional half-turn with a wrench or pliers *(Step 1)*.

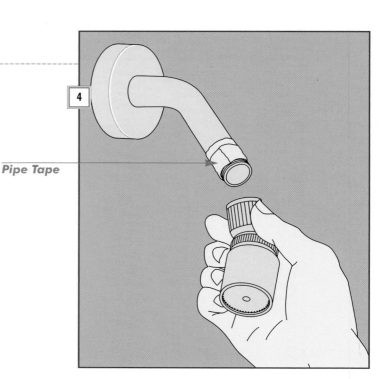

Pipe Tape

WATER-SAVING SHOWER HEADS

If you use a shower head that is more than several years old, it's probably a high-flow model that wastes money every time you turn on the hot water. A low-flow shower head is the answer. Today's models offer a comfortable shower while using considerably less hot water. Best of all, they are inexpensive and easy to install.

Some models operate by restricting the flow of water through the head (and thus increasing the pressure). Others, such as the one shown here, operate on the principle that the best way to save water is not to use it in the first place. These heads have a knob or switch on one side that allows you to instantly stop the flow of water when you shampoo your hair or shave. A quick flick of the knob turns the water back on (and at the same temperature, too).

Septic Systems

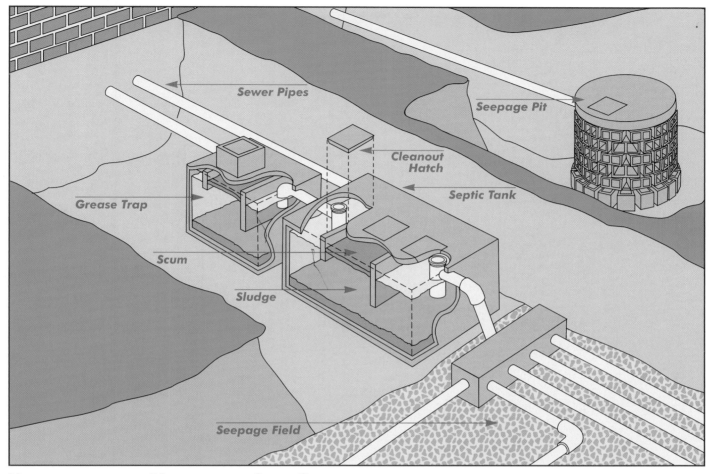

Sewer Pipes

Seepage Pit

Cleanout Hatch

Grease Trap

Septic Tank

Scum

Sludge

Seepage Field

UNDERSTANDING AND MAINTAINING A SEPTIC SYSTEM

Household sewage—99% of which is water—moves from the house through the sewer line to a grease trap, and then to the septic tank for treatment. (Toilet waste drains directly into the tank, bypassing the trap.) At the tank, it is divided into four components: sludge and scum, which are trapped in the tank; gas, which escapes through the sewer pipe and out the house vent; and liquid sewage, which flows to a seepage field beneath the lawn, where it filters harmlessly into the ground. If your pipes back up, the tank might be full. Have it inspected and cleaned at least as often as required by the local health department, usually every two to five years, but more often if the kitchen sink has a garbage disposer. Some septic systems also include a seepage pit to handle "gray water" from sinks, bathtubs, and clothes washers, which reduces loads on the main septic system.

When part of the lawn remains soggy even in dry weather, or household drains are sluggish or smell of sewage, have a professional service the septic system. The tank may need to be pumped out, the grease trap may require cleaning, or the outlet pipe could be blocked.

CAUTION

To function properly, a septic tank requires an environment in which bacteria can attack and break down the wastes. This environment can be disrupted, however, when strong chemicals are poured into the drains. It is always best to avoid the overuse of chemicals such as bleach, toilet bowl cleaners, and drain openers. If you are not certain about what's permissible to flush or pour down a drain, ask a qualified professional beforehand.

FIX IT: Pipes and Fittings

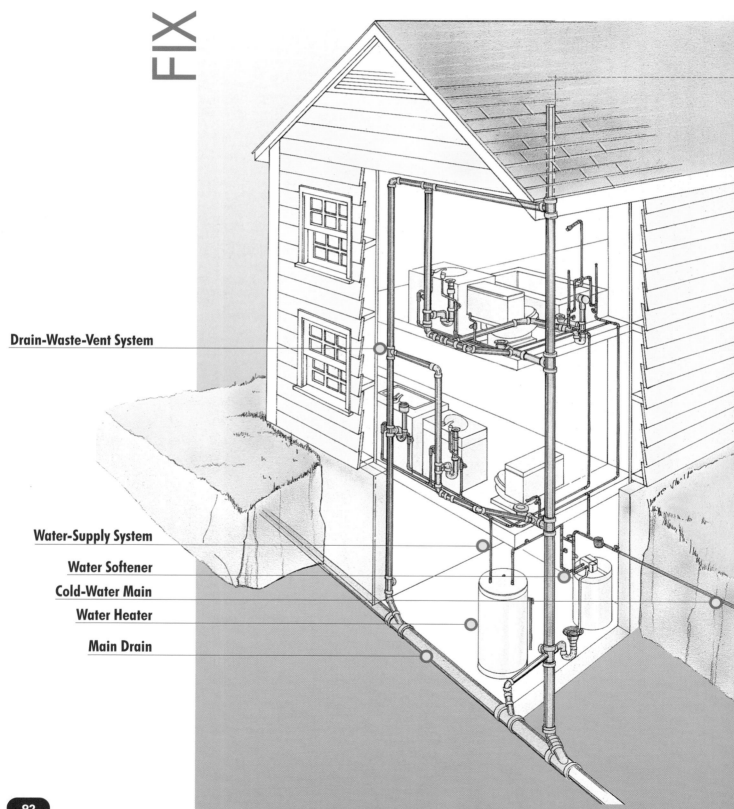

Drain-Waste-Vent System

Water-Supply System

Water Softener

Cold-Water Main

Water Heater

Main Drain

Chapter 6

How They Work

Domestic plumbing consists of three basic systems: supply lines, fixtures, and drainpipes. Although it may seem a puzzle of pipes and fittings, the system and its various components work in a logical way.

The supply system consists of copper, steel, or plastic pipes, fittings, and valves that carry potable water throughout the house. From the cold-water main, one supply pipe branches off to the water heater (or, as here, a water softener and then to the water heater) to begin a run called the hot-water main. Secondary branches of hot and cold water run through walls and between floors to the various fixtures.

Plumbing fixtures include sinks, toilets, and anything else connected to the supply and drainage systems.

Drainpipes comprise the drain-waste-vent (DWV) system. While it is the least visible part of the system, it is the part most strictly regulated by plumbing codes. The system depends on gravity, not pressure, to carry waste water out of the house. Each fixture is connected to a drainpipe by a trap filled with water that prevents harmful sewer gas from entering the home.

Contents

Troubleshooting

Problem	Solution
• **Pinhole leak in pipe**	Patch hole temporarily until a permanent repair can be made **99** •
• **Fitting or joint loose**	Resolder copper joint **103** • Tighten or replace fitting **114** •
• **Copper pipe cracked or corroded**	Replace with copper **100** • Replace with CPVC **104** •
• **Galvanized steel pipe cracked or corroded**	Replace with steel **106** • Replace with copper **108** •
• **CPVC pipe cracked**	Replace with CPVC **110** •
• **Cast-iron drainpipe leaks**	Tighten hose clamps, if any; otherwise call plumber **112** •
• **Plastic drainpipe cracked**	Replace damaged section **111** •
• **Globe valve leaks**	Tighten or replace fittings **114** • Disassemble and clean valve **115** •
• **Globe valve won't close**	Disassemble and clean valve, or replace it **115** •
• **Pipe or faucet noisy**	Install expansion tank to control water hammer **117** •

Before You Start

When a piping repair is called for, the first decision concerns materials: should you match what's already in place or switch to something different.

COPPER, STEEL, AND PLASTIC

New methods and materials put the art of pipe fitting well within reach of the homeowner. Choosing the most appropriate pipes for a given situation will make the repair process as easy as possible. In most cases, damaged piping is repaired or replaced with the same material, but there are reasons for considering other alternatives.

Copper, today's standard for supply pipes, is relatively easy to work with and readily available in rigid, 10-foot lengths. Two thicknesses are available for home use: medium-thick Type L; and low-cost but thin-walled Type M, which is suitable for most repairs.

Galvanized steel is the strongest plumbing material, but it must be joined with threaded fittings and is prone to corrosion. While steel pipe normally comes in 21-foot lengths, most plumbing supply stores will cut and thread a shorter piece for you.

Supply pipe made from chlorinated polyvinyl chloride (CPVC) is very light, inexpensive, and the easiest of all to use; however, local codes don't always permit its use. Drainage and vent pipes are typically polyvinyl chloride (PVC) or acrylonitrile butadiene styrene (ABS); both are quite easy to work with.

TOOLS

Ruler or steel tape
Dividers or calipers
Propane torch and flux brush
Screwdriver
Safety goggles
Heavy work gloves
Fire extinguisher
Keyhole saw
Pipe wrench and pliers
Tube cutter
Hacksaw
Miter box
Sharp knife

MATERIALS

Pipe
Pipe fittings
Emery cloth
Solder
Flux
Pipe tape
Electrical tape
Old bicycle inner tube
Hose clamps
Clean rags
PVC or CPVC primer and cement
ABS cement

Chapter 6

Types of Fittings

Supply Fittings

Nipple
Joins threaded fittings that are close. Often used to complete a run.

T
Joins a 90-degree branch run to a straight run of pipe.

Coupling
Joins lengths of pipe running in the same direction. Pipe ends bottom out against a shoulder inside a standard coupling; a slip coupling has no shoulder.

Street elbow
Attaches to another fitting to change pipe direction.

Union
Joins threaded pipe, usually galvanized steel, so that the run can be disassembled.

Plug
Closes an unused opening in a pipe or fitting. Also used to temporarily plug an open pipe during a repair if the water must be turned on.

Elbow
Changes the direction of a pipe run. Available in 45- and 90-degree bends.

Transition Fittings

Dielectric union
Joins copper pipe to steel. Prevents corrosion between the different metals. The threaded end screws onto the steel pipe; the brass end is soldered onto the copper pipe.

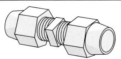

Compression fitting
Joins copper or CPVC supply lines without cement or solder.

CPVC-to-unthreaded pipe adapter
Compression side of adapter clamps to copper pipe; other end is cemented to CPVC pipe.

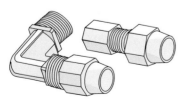

Threaded adapters
Used to join copper or CPVC pipe to threaded pipe. When buying adapters, specify whether they will be used for hot or cold water-supply pipes.

CPVC-to-threaded pipe adapter
Threaded side of adapter screws onto steel or brass pipe; other end is cemented to CPVC pipe.

Drain Fittings

Quarter bend (90-degree elbow)
Changes direction of PVC or ABS drainpipe.

Sanitary Y with cleanout
Joins two lengths of drainpipe; threaded cleanout allows access to the pipe for cleaning.

Reducer
Connects drainpipes of different diameters.

Closet flange and bend
Together, they connect the toilet to its branch drain and anchor the toilet to the bathroom floor.

Hubless fitting
Joins hubless cast-iron, ABS, or PVC drainpipe without caulking or cement. A tight-fitting neoprene sleeve is held in place by a stainless steel collar and clamps.

Y branch
Joins a branch pipe at an angle to a straight pipe run.

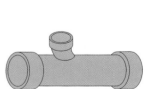

Reducing T-Y
Joins a small pipe at a 90-degree angle to a larger-diameter pipe.

Measuring Pipe Diameter

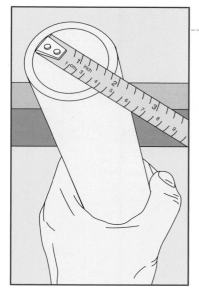

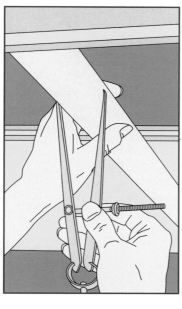

AT THE END OF A PIPE

Always choose replacement pipe and fittings based on inside diameter.

- Hold a ruler or steel tape across the widest part of a cut pipe *(left)*.

- To determine the inside diameter (ID), or nominal size of the pipe, measure from one inner wall to the other.

IN THE MIDDLE OF A RUN

- If the pipe is part of a run, first span the outside diameter (OD) using dividers *(left)*, calipers, or a C-clamp, then measure the result with a ruler.

- Take several readings and average them, then refer to the chart on page 98 to find the inside diameter (ID).

Measuring Pipe Length

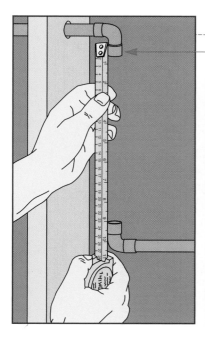

Shoulder

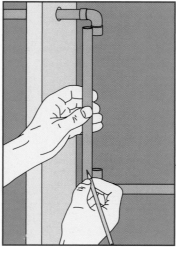

MEASURING A GAP

- After cutting out or otherwise removing a damaged pipe, buy a replacement pipe that is several inches longer than the gap.

- Place new fittings on the ends of the sound pipes as needed. With a steel tape, measure between the shoulders of the fittings, then cut the new pipe to this length *(far left)*.

SCRIBING A PIPE

- An alternative method is to hold the replacement pipe at the gap and mark the exact length, including the depth of the fittings, with a pencil *(left)*, then cut the pipe.

CALCULATING PIPE DIMENSIONS

TYPE OF PIPE	OUTSIDE DIAMETER	INSIDE DIAMETER	DEPTH OF FITTING
COPPER			
	3/8 in.	1/4 in.	5/16 in.
	1/2 in.	3/8 in.	3/8 in.
	5/8 in.	1/2 in.	1/2 in.
	7/8 in.	3/4 in.	3/4 in.
	1 1/8 in.	1 in.	15/16 in.
	1 3/8 in.	1 1/4 in.	1 in.
	1 5/8 in.	1 1/2 in.	1 1/8 in.
GALVANIZED STEEL			
	3/8 in.	1/8 in.	1/4 in.
	1/2 in.	1/4 in.	3/8 in.
	5/8 in.	3/8 in.	3/8 in.
	3/4 in.	1/2 in.	1/2 in.
	1 in.	3/4 in.	9/16 in.
	1 1/4 in.	1 in.	11/16 in.
	1 1/2 in.	1 1/4 in.	11/16 in.
	1 3/4 in.	1 1/2 in.	11/16 in.
	2 1/4 in.	2 in.	3/4 in.
PLASTIC			
	7/8 in.	1/2 in.	1/2 in.
	1 1/8 in.	3/4 in.	5/8 in.
	1 3/8 in.	1 in.	3/4 in.
	1 5/8 in.	1 1/4 in.	11/16 in.
	1 7/8 in.	1 1/2 in.	11/16 in.
	2 3/8 in.	2 in.	3/4 in.
	3 3/8 in.	3 in.	1 1/2 in.
	4 3/8 in.	4 in.	1 3/4 in.

READING THE CHART

Always choose replacement pipe with the same inside diameter as the old pipe. For example: You have removed a damaged section of copper pipe with a 7/8-inch outside diameter and a 3/4-inch inside diameter, and have chosen to replace it with plastic pipe. Buy plastic pipe with a 3/4-inch inside diameter (its nominal size), but note that the depth of the plastic fittings will be 5/8 inch, not 3/4 inch as with copper fittings. Be sure to account for this difference when you cut the pipe.

Emergency Pipe Repairs

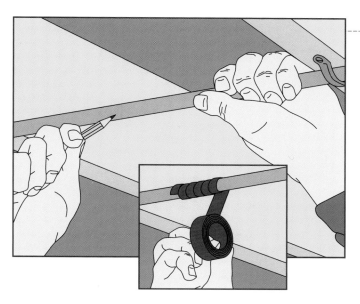

A PENCIL-TIP STOPPER

• As a temporary repair for a leak, turn off the water supply at the main shutoff valve, jam a pencil point into the hole *(left),* and break it off. A toothpick may also work, but for steel pipes, the graphite in the pencil tip will better seal the leak.

• Dry the pipe and tightly stretch two or three layers of plastic electrician's tape around the pipe *(inset).* The tape should extend three inches on each side of the leak; overlap each turn by half. For a more permanent solution, replace the damaged section of pipe.

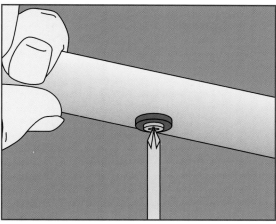

A SCREW-AND-WASHER PLUG

• Turn off the water supply at the main shutoff valve. Place a faucet washer (or any convenient rubber washer) on a sheet-metal screw short enough not to pierce the other side of the pipe.

• Set the tip of the screw in the leak and tighten the screw gradually *(left).* In copper pipe, take special care not to overtighten the screw.

• Make a permanent pipe repair as soon as possible.

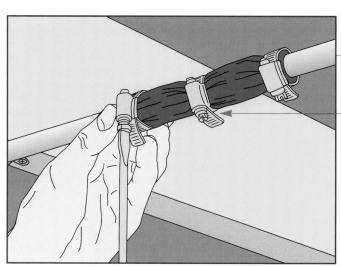

Hose Clamp

AN INNER-TUBE SLEEVE

• For a slightly larger crack or puncture, close the main shutoff valve and drain the supply line. Wrap the pipe with an old bicycle inner tube *(left)* and secure it with hose clamps. At least one clamp should be directly over the damaged section.

• Turn the water back on, slowly at first, to test for leaks. Replace the damaged section as soon as possible.

Replacing Copper Pipe

1. CUTTING COPPER PIPE

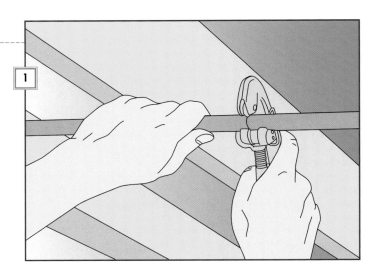

- Close the main shutoff valve and drain the supply lines by draining the system through the water heater *(page 128)*.

- Fit a tube cutter around the pipe next to the break. Turn the knob clockwise until the cutting disk bites into the pipe *(right)*.

- Rotate the cutter once around the pipe, then tighten the knob and rotate again. Continue tightening and turning the tube cutter until the pipe is nearly severed, then snap it apart with your hands.

- Loosen the knob, slide the cutter down the pipe, and cut the pipe on the other side of the break.

2. DEBURRING THE CUT END

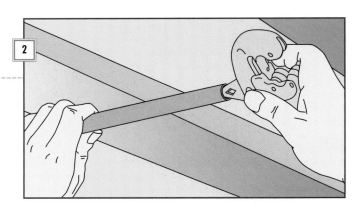

- Use the triangular blade attached to the cutter (or a round metal file) to ream out the burrs inside the old and new pipes.

FIXING A LEAKY JOINT

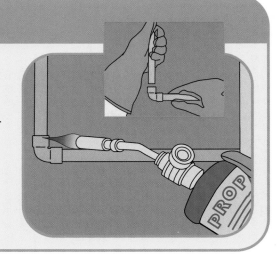

The cure for a leaking joint is to take it apart and resolder it. Wear safety goggles and heavy work gloves, protect flammable materials near the joint with a fireproof shield, and have a dry-chemical fire extinguisher on hand. Turn off the water and drain the pipe (you may have to drain the entire system; see Step 1 above). Play the flame of a propane torch over the fitting until the solder melts *(right)*. Pull the pipe from the fitting *(inset)*, then resolder the joint *(page 103)*.

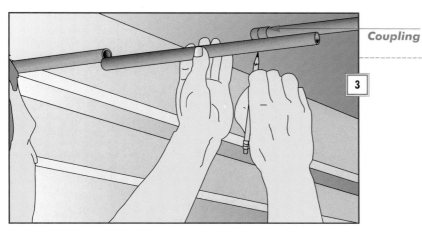

Coupling

3

3. MEASURING PIPE

- Fit standard couplings on the ends of the old pipe, hold the new pipe against the gap and mark it at the coupling ridges. (If you are using slip couplings, mark the new pipe to fill the gap completely.)

- Cut the replacement pipe at the mark with a tube cutter.

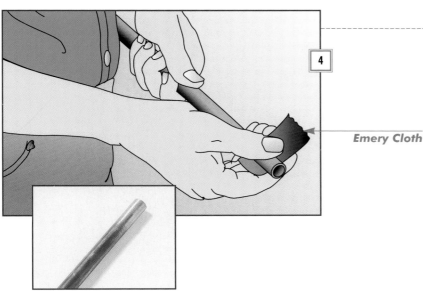

4

Emery Cloth

4. PREPARING THE JOINTS

- Rub the inside of the couplings and the ends of the old and new pipes with emery cloth until they are bright and shiny *(left)*.

- Remove any grit left on the surfaces with a clean, dry cloth.

5

Flux Brush

5. BRUSHING JOINTS WITH FLUX

- Using a small, stiff brush, spread a thick and even coat of soldering flux on all of the cleaned pipe surfaces *(left)*.

- Brush a small amount of flux inside the couplings.

6. FITTING THE COUPLINGS

- Dry the insides of the standing pipes near the ends.

- Fit a coupling onto the ends of the standing pipes *(right)*; each coupling should bottom out on the pipe. (If you are using slip couplings, put one on each standing pipe and slide the couplings into place.)

- Give each coupling a quarter-turn to evenly spread the flux.

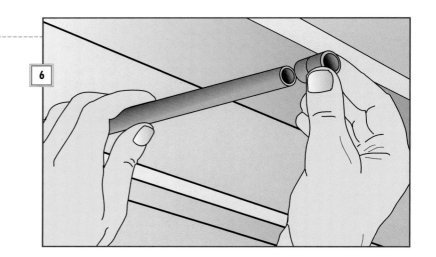

7. FITTING THE REPLACEMENT PIPE

- Insert the replacement pipe into one coupling.

- Gently pull the pipes toward you *(right)* until you can slip the other end into the second coupling. (If you are using slip couplings, hold the new pipe in place, then slide each coupling over the joints.) Give the new pipe a quarter-turn to evenly distribute the flux.

- Do not move the pipe or the couplings excessively or you will remove flux from the joints.

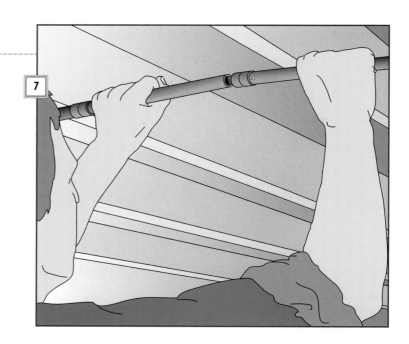

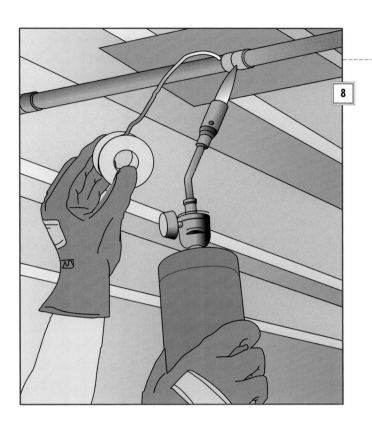

8. SOLDERING THE JOINT

• Wear safety glasses and leather work gloves, and protect flammable materials with a fireproof shield.

• Light the torch and play the flame over the fitting and nearby pipe, heating them as evenly as possible.

• Touch the tip of the wire solder to the joint until it melts into the fitting *(left)*. As soon as the solder enters the joint, direct the flame down the coupling so that the solder flows internally. Do not let the flame touch the solder. When the joint is properly heated, capillary action draws molten solder into the fitting to seal the connection.

• Feed solder into the joint until a bead of metal appears around the edge.

COMPARING SOLDER JOINTS

A good solder joint *(far right)* is smooth and shiny; a poor one *(near right)* looks burned and lumpy, and has tiny pinholes in the solder that will probably leak sooner or later.

When soldering with a propane torch, remember that the tip of the inner (darker) cone of flame is the hottest. Keep the torch moving evenly over the joint; if you apply too much heat, flux will burn away and the copper will darken. If this happens, clean the joint and flux it again. The perfect joint results when the heat is maintained at just the point where the solder flows evenly.

Practice your soldering technique on some spare pipe and fittings. To check a practice joint, hacksaw it apart after it cools. Solder should completely fill the gap between the pipe and the fitting.

Substituting CPVC for Copper

1. FITTING THE NEW SECTION

• Prepare to be without water for as long as it takes CPVC solvent cement to cure (two hours at temperatures above 60°F), or use a compression fitting *(page 96)*. Close the main shutoff valve; drain supply lines through the water heater *(page 128)*.

• Cut the broken section of copper pipe with a tube cutter or hacksaw. Deburr the ends of the standing pipes *(page 100)*.

• Hold the CPVC pipe against the gap and mark it with a pencil *(right)*.

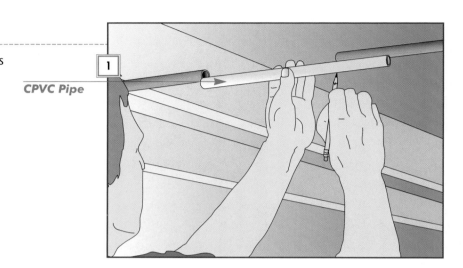

CPVC Pipe

2. CUTTING THE REPLACEMENT PIPE

• Place the CPVC pipe in a miter box, lining up the pencil mark on the pipe with the slotted saw guides.

• Brace the pipe with one hand *(right),* and saw through the mark with a hacksaw. If you are cutting several pipes, a plastic tube cutter will speed the job.

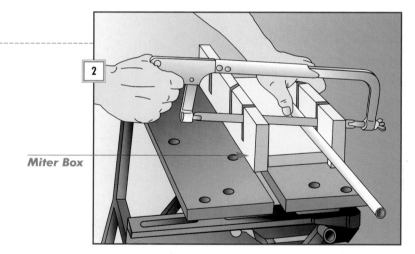

Miter Box

3. DEBURRING AND BEVELING CPVC

• The ends of a sawed-off CPVC pipe must be deburred and beveled. Using a sharp knife, trim the inside edge to aid water flow and the outside edge *(right)* to improve the welding action of the cement.

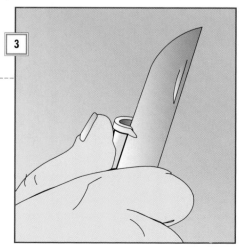

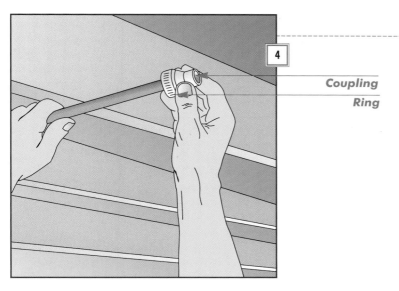

Coupling

Ring

4. Fitting the adapter couplings

- Loosen the knurled ring on two adapter couplings and push them onto the copper pipes until the pipe ends bottom out inside the coupling sockets *(left)*. To make the job easier, clean the edges of the copper pipes with emery cloth. If you mark the pipes about 1 1/2 inches from the ends, you will be able to tell if they are properly seated.

- Hand-tighten both couplings.

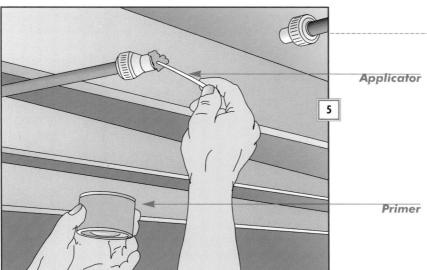

Applicator

Primer

5. Priming and cementing the joints

- With an applicator or clean cloth, apply a coat of primer to the ends of the CPVC pipes and the adapter couplings *(left)*. Let the primer dry for a few seconds.

- With another applicator, apply a liberal coat of CPVC cement as deep as the coupling sockets to the outside ends of the CPVC pipe. Then, working quickly—the cement sets in less than 30 seconds—apply a light coat of cement inside the coupling sockets.

CAUTION: Cement fumes are toxic and highly flammable. Ventilate the work area and do not smoke or use an open flame.

6. Fitting the replacement pipe

- Immediately push one end of the CPVC pipe into a coupling. Pull the free ends toward you until it is possible to slip the CPVC pipe into the second coupling *(left)*. Give the CPVC pipe a quarter-turn to evenly distribute the cement inside the sockets.

- Wipe away any excess cement with a clean, dry cloth.

- Turn on the water. Tighten the fittings a quarter-turn if you spot any leaks.

Fixing Steel Pipe

1. CUTTING OUT A DEFECTIVE PIPE

● Close the main shutoff valve and drain the supply lines by draining the water heater *(page 128)*.

● Once in place, a threaded pipe cannot be unscrewed as one piece, since loosening it at one fitting will tighten it at the other. Look for a union near the damaged section and, if there is one, unscrew the pipe from it. Unscrew the other end from its coupling.

● If there is no union, first cut through the pipe with a fine-tooth hacksaw or mini-hacksaw *(right)*, then unthread the two pieces from their couplings *(below)*.

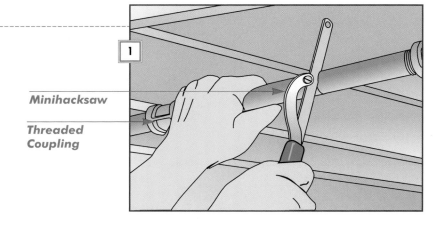

Minihacksaw

Threaded Coupling

2. REMOVING THE DEFECTIVE PIPE

● To unthread pipe from a union or coupling, grip the coupling with one wrench and turn the pipe with another so that the rest of the run will not be twisted or strained. Turn the jaws of the wrenches to face the direction in which force is applied.

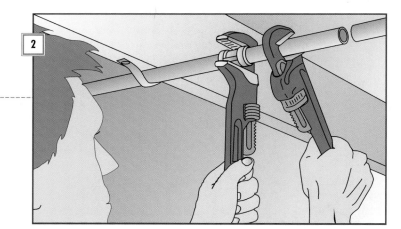

ELECTRICAL GROUNDING AND PIPE REPAIR

In a safe house-wiring system, an unbroken electrical path must link all electrical components to the main electrical panel, which in turn is connected by a wire to a metal rod driven several feet into the ground outside the house and to metal, cold-water pipes that exit the house into the ground. The system discharges into the ground abnormal bursts of electrical current.

It's possible to interrupt this grounding circuit if you insert plastic pipe or a dielectric union in the piping run. To maintain a proper ground, a ground clamp should be attached to the metal pipes on both sides of the repair, then connected with copper jumper wire. Check with local building code officials for specific requirements.

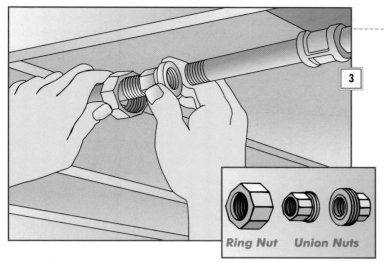

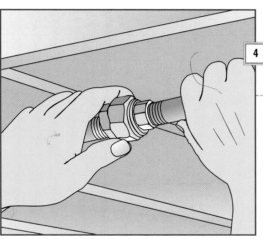

Ring Nut   Union Nuts

3. PREPARING THE UNION

• Buy two lengths of new pipe and a union whose combined length, when screwed together, is the same as the broken section. Have the pipes threaded.

• Wind 1 1/2 turns of pipe tape clockwise over the threads of the new pipes, tightly enough so that the threads show through.

• Thread a pipe into the nearest coupling using a double-wrench technique like that shown in Step 1. Then thread the other pipe into its fitting.

• Disassemble the union. Slide its ring nut onto one of the pipes *(left)*, then screw the union nuts onto the pipe ends, deflecting the pipes as needed.

4. CONNECTING THE UNION

• Slide the ring nut to the center of the union and screw it onto the exposed threads of the union nut, joining the two pipes *(left)*.

• Grip the exposed union nut with one wrench and tighten the ring nut with a second wrench.

Ron's TRADE SECRETS

MAKING GOOD CONNECTIONS

Galvanized pipe relies on threaded connections, and those in turn rely on threads that are clean and undamaged. I always clean the outside and inside threads before joining parts. Small wire brushes *(right)* work great for getting rid of dirt and scale. Brushes with stainless steel bristles are best for removing tough deposits from hard metals such as galvanized pipe. Brass brushes are best for brass threads, or for lightly cleaning galvanized pipes. And I always use Teflon™ tape or a brush-on pipe sealant on threaded connections. Either one of these will act as a lubricant ensuring the threads will form a tight seal.

Replacing Steel with Copper

1. MEASURING THE PIPE

- Close the main shutoff valve and drain the supply lines by draining the water heater *(page 128)*.

- Unscrew the damaged section of galvanized pipe *(page 106)*, and remove the couplings.

- Measure the amount of copper replacement pipe you will need. Be sure to allow for the fittings *(below)*.

Galvanized Pipe

Copper Repair Pipe

1

2. FITTING THE UNION

- Screw the steel spigots of the dielectric unions onto the steel pipe ends, and apply pipe tape to exposed threads.

- Support the replacement section by fastening it to a joist with a pipe anchor.

- Slip the ring nuts and plastic collars past the ends of the copper pipe *(right)*.

2

Collar

Ring Nut

Spigot

STOPPING DIELECTRIC CORROSION

When dissimilar metals touch, an electrolytic chemical reaction takes place between them. The intensity of the reaction depends in part on the metals involved. Copper and steel, for instance, react vigorously enough to eat away at any joint where the two metals touch, eventually causing it to leak.

To prevent this outcome, special, dielectric unions *(right)* are required to join pipes of copper and galvanized steel. Dielectric unions differ from ordinary unions *(page 107)* in having a plastic collar and a rubber washer that prevent the two metals from touching.

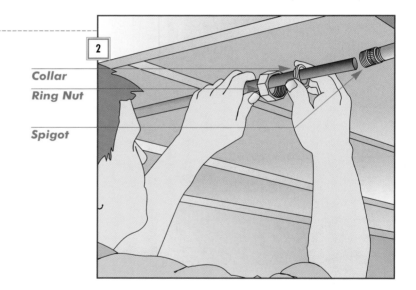

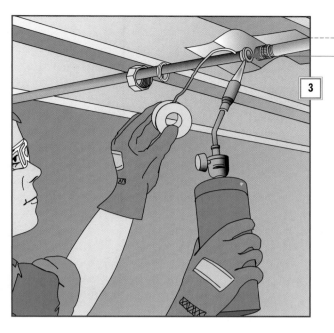

Spigot

3

3. SOLDERING THE BRASS FITTINGS

● Wear safety glasses and work gloves, and protect flammable materials with a fire-proof shield.

● With an emery cloth, burnish the brass shoulders of the two dielectric unions and both ends of the copper replacement pipe.

● Apply flux to the inside of the brass shoulders and the outside of the copper pipes *(page 101)*. Slip the shoulders onto the ends of the pipes and give them a quarter-turn to evenly distribute the flux.

● Solder the shoulders to the pipes *(left)* and allow them to cool before completing the repair.

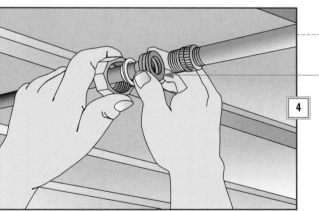

Washer

4

4. FITTING THE REPLACEMENT PIPE

● Place a rubber washer against face of the soldered brass shoulders, then slide the collar and ring nut up to the spigot and tighten the nut by hand.

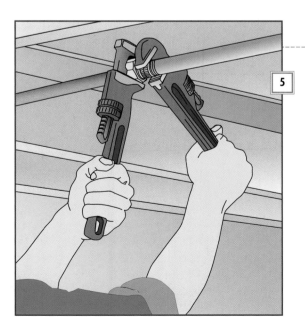

5

5. TIGHTENING THE UNION

● Grip the spigot with one wrench and tighten the ring nut one half-turn with a second wrench *(left)*.

● Repeat this procedure to assemble and tighten the other union.

CAUTION: If you are repairing a cold-water pipe, assume that the dielectric unions have interrupted your home's grounding circuit. Install a jumper wire *(page 106)*.

Repairs in CPVC Pipe

1. PREPARING THE JOINT

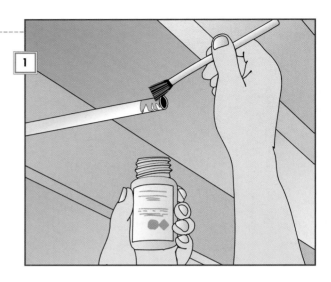

- Close the main shutoff valve, drain the supply lines through the water heater *(page 128)*, and cut out the damaged section of pipe.

- Cut the new pipe to fit *(page 97)*, then deburr and bevel all of the pipe ends *(page 104)*.

- Clean the ends of two coupling sockets and the existing pipes with CPVC primer. Wait a few seconds for it to dry, then apply CPVC cement to the couplings and pipe ends *(right)*.

- Push the couplings onto the pipes; give them a quarter-turn to spread the cement. Hold the pieces together for 10 seconds.

CAUTION: Cement fumes are toxic and highly flammable. Ventilate the work area and do not smoke or use an open flame.

2. INSERTING THE REPLACEMENT PIPE

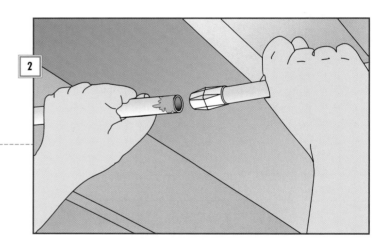

- Clean and prime the exposed coupling sockets and both ends of the replacement pipe, then apply cement.

- Push one end of the replacement pipe into a coupling, then gently bend the pipes toward you until you are able to slip the replacement into the other coupling *(right)*.

- Give the pipe a quarter-turn and wipe off excess cement with a clean, dry cloth. Do not run water in the pipe until the CPVC has cured (about two hours at temperatures above 60°F).

ASSEMBLING RUNS

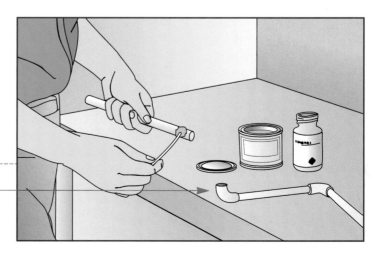

- If the repair involves a number of fittings and lengths of pipe, it's usually easier to assemble the pieces into "runs" before lifting them into place.

Run

A Fix for Plastic Drainpipe

Paper

Keyhole Saw

Slip Coupling

1. REMOVING THE DAMAGED SECTION

• Do not run water or flush toilets in the house during this repair.

• Wrap a sheet of paper around the pipe as a saw guide and cut squarely through the pipe with a keyhole saw or handsaw. Repeat at the other end of the damaged section.

• As you proceed, stuff newspapers or paper towels into the standing pipes to block sewer gas.

2. CUTTING THE REPLACEMENT PIPE

• Fit a slip coupling over each standing pipe. Carefully measure the gap *(left)* and transfer the measurement to the new pipe.

• Cut the replacement section to length, using a handsaw and miter box to make a square cut.

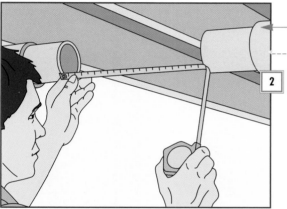

3. APPLYING CEMENT

• With a sharp knife, deburr and bevel the pipe ends *(page 104)*. Clean the ends of PVC and ABS pipes (but prime only PVC pipes).

• Apply a thick coat of cement half the width of the slip coupling to all four pipe ends *(left)*.

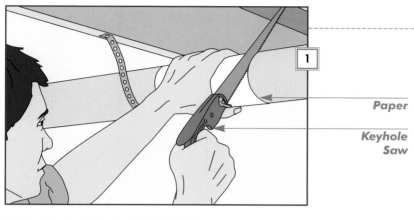

PVC

4. Fitting the replacement pipe

● Working quickly, lift the new pipe into place and slip one coupling, then the other, over its joint. Each coupling must cover an equal amount of old and new pipe.

● Give each coupling a quarter-turn to spread the cement *(right)*, then wipe away excess cement with a clean cloth.

● Allow the joint to cure (at least two hours), then run water through the drainpipe. If it leaks, the couplings were not properly cemented. Cut out the replacement pipe and fittings and start again with new parts.

CAST-IRON DRAINPIPE

Cast-iron drainpipe, rarely found in new homes but common in older ones, is a durable product that muffles the sound of water rushing through the piping. Repairing damaged cast-iron pipe, however, is heavy work that many homeowners leave to a plumber. The job begins with cutting out the damaged section of pipe, either with a circular saw equipped with a special blade, or with a soil-pipe cutter—a tool similar in principle to a tube cutter *(page 100)*. Either tool can be rented.

A length of replacement pipe, cut to bridge the gap cut in the existing drainpipe, can then be secured to the standing pipes with hubless fittings, which are neoprene sleeves that work like slip fittings. Once the new pipe and hubless fittings are in place, each fitting is secured by four stainless steel hose clamps *(right)*.

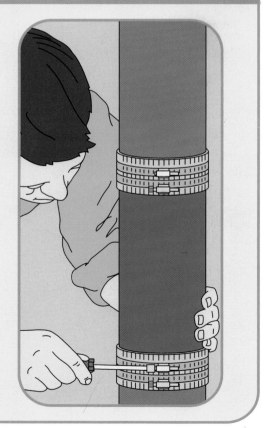

Shutoff Valves

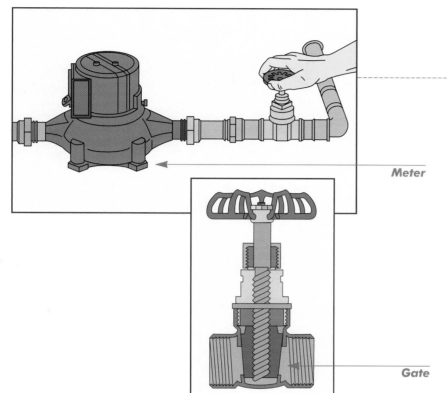

Meter

Gate

LOCATING THE MAIN SHUTOFF VALVE

The main shutoff valve controls the flow of water to the entire house. It is usually a gate valve *(inset),* and is typically found in an upright position that prevents sediment from collecting in the closing mechanism. As the handle is turned clockwise, the internal gate lowers to block the flow of water.

• Look for the valve near the water meter in the basement, utility room, or crawl space. It may even be located outside in mild climates.

• Turn the handle clockwise to shut off the flow of water.

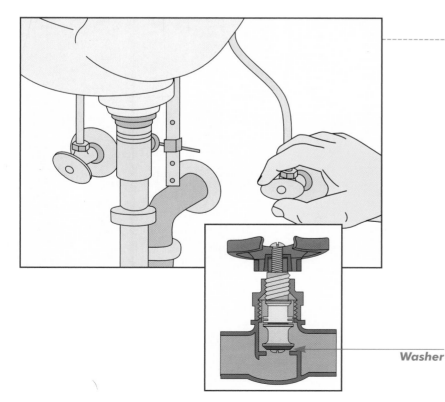

Washer

LOCATING FIXTURE SHUTOFF VALVES

Fixture shutoff valves cut the water supply to an individual fixture or appliance without affecting others. Most are globe valves *(inset).* When the handle turns, it forces a rubber washer or disc against a matching seat, blocking the flow of water. Partitions within the valve body slow the flow of incoming water, reducing water volume through the supply line.

• Look beneath the appliance or fixture on the water-supply piping. However, fixtures do not always have individual shutoff valves, especially in older houses. Valves are sometimes installed on branch lines instead; look in the basement on pipes that serve the device you wish to turn off.

• Turn the handle clockwise to shut off the water.

Stopping Supply-Tube Leaks

1. TIGHTENING THE COUPLING NUT

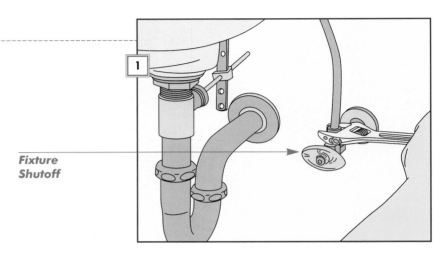

* Using an adjustable wrench, tighten the coupling nut on the fixture shutoff *(right)*, as well as the packing nut behind the handle. Some plastic nuts may need only hand-tightening.

* Turn on the water. If a joint leaks, tighten the coupling nut an additional half-turn with a wrench and test again. If the faucet has an aerator, remove it and open the faucet to clear debris that may have been loosened by the repair.

Fixture Shutoff

2. REPLACING A FAULTY COUPLING NUT

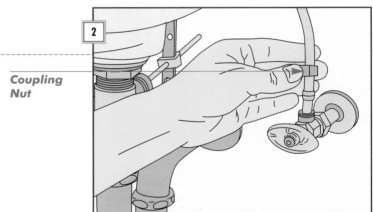

* If the nut still leaks, it may be faulty. Remove it, along with the brass compression ring beneath (you may have to cut the ring off). Slip a new nut and ring onto the free end of the supply tube, then push the tube into the valve outlet as far as it will go.

Coupling Nut

* Lower the compression ring onto the joint, making sure that it is squarely aligned, then slide the coupling nut *(right)* over the fitting and screw it down by hand.

* Turn on the water. If the joint drips, gradually tighten the nut with a wrench until the leak stops.

3. CHECKING THE FIXTURE FITTINGS

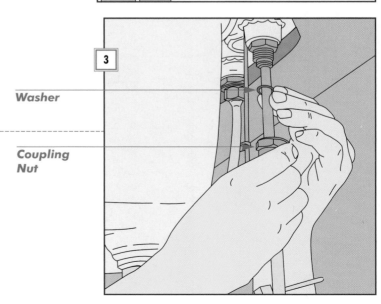

* If the leaking nut is on the upper end of the supply tube and is difficult to reach, tighten it with a basin wrench *(page 28)*, or replace it. This coupling nut may incorporate a washer as part of the assembly.

Washer

Coupling Nut

Dealing with Globe Valves

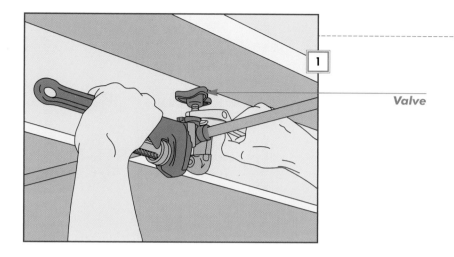

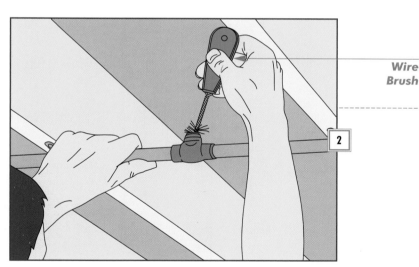

Valve

Wire Brush

Metal Heat Shield

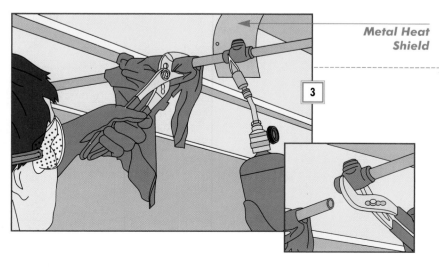

1. REMOVING THE VALVE STEM

- Close the main shutoff valve and drain the supply lines through the water heater *(page 128)*.

- Grip the valve body with a pipe wrench and hold it steady while loosening the bonnet nut with a wrench *(left)*. If it is stubborn, apply penetrating oil and wait 15 minutes before trying again. Pull the valve stem from the faucet body.

- If water was seeping from the stem and handle, clean the valve body in place *(Step 2)*. If the valve was leaking badly or did not close, replace it as shown beginning with Step 3.

2. CLEANING THE VALVE BODY

- Remove mineral deposits and sediment inside the valve body with a wire brush designed for cleaning copper fittings *(left)*.

- Scour the valve opening, then bend the wire shaft of the brush with a pair of pliers to reach farther inside the valve.

- Replace the washer, O-ring, or packing under the bonnet nut, and the washer at the end of the stem. Reassemble the faucet.

3. REMOVING THE VALVE

- Desolder the valve as you would a fitting in a copper pipe *(page 100)*. The valve's thick body heats slowly. Open any other valves or faucets on the line, and wrap a damp rag around the adjacent pipes to cool them.

- As the solder melts, pull the pipe straight out of the valve without bending it *(inset)*.

4. Soldering the New Valve

● Brighten the inside of the valve and the ends of the pipes with emery cloth. Remove any grit with a clean, dry cloth.

● With a small brush, spread a thick, even coat of soldering flux on all of the pipe surfaces. Brush a small amount of flux inside the valve.

● Solder the new valve body *(right)* into the pipe run as shown on page 103.

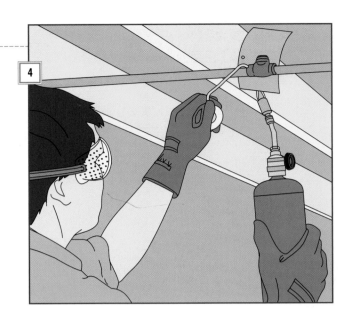

5. Reassembling the Valve

● When the valve body cools, thread the stem into the valve opening *(right)*. Grip the valve body with one wrench and tighten the bonnet with another.

● To install a threaded valve, screw it to the pipe by hand, then grip the pipe with one wrench and tighten the valve with a second wrench. Thread a new piece of pipe into the other side of the valve and add a union to complete the run *(page 107)*.

Stem

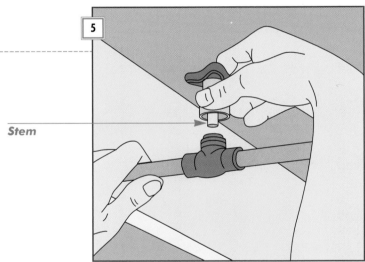

HIDDEN SHUTOFF VALVES

Shutoff valves for sinks and toilets are usually pretty easy to pinpoint, but the locations of other shutoffs may present a mystery. When looking for a shutoff valve that's not where you might expect it to be, try to trace the water-supply lines from the fixture or appliance to the main. This bit of detective work may uncover a dishwasher shutoff in the basement, directly under the appliance. The shutoff for an outdoor spigot may be tucked between floor joists above a basement or crawl space. Others may be hiding behind access panels or in the dropped ceiling of a basement recreation room.

Banishing Water Hammer

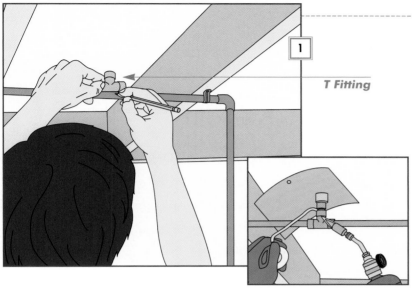

T Fitting

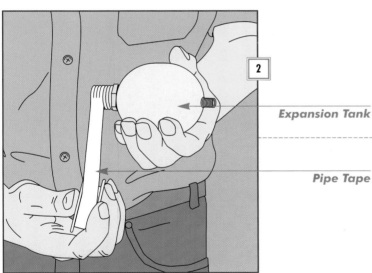

Expansion Tank

Pipe Tape

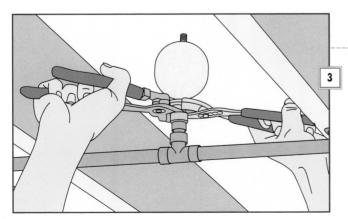

1. INSTALLING THE T AND ADAPTER

Water hammer is a banging noise caused by vibrating pipes, and is amplified when the pipes are in contact with the framing or walls. You can greatly reduce or eliminate this annoyance by installing an air-filled expansion tank on a vibrating pipe.

• Close the main shutoff valve and drain the supply lines. From the noisy faucet, trace the supply pipe to the nearest accessible location and mark the pipe for a copper T fitting *(left)*. Cut out a section just long enough to accept the T.

• After soldering the T in the supply line *(inset)*, solder a 2-inch copper nipple into the top of the T, then solder a female-threaded brass adapter to the nipple. The T and adapter can be installed upside-down if clearance is tight.

2. PREPARING THE EXPANSION TANK

• Wrap pipe tape around the threads of the expansion tank *(left)*.

• Screw the tank into the brass adapter by hand.

3. INSTALLING THE EXPANSION TANK

• Grip the adapter with a pair of channel-joint pliers and tighten the expansion tank with a second pair or a wrench *(left)*.

• Support the pipe on each side of the expansion tank with pipe hangers.

• Turn on the water, slowly at first. If there are any leaks, tighten the tank another half-turn.

FIX IT: Gas Water Heaters

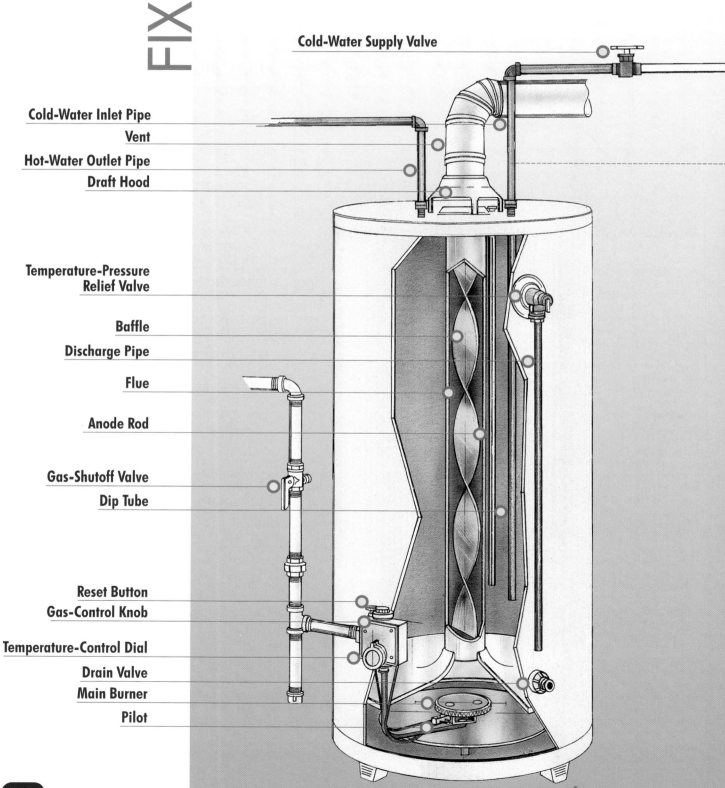

Cold-Water Supply Valve

Cold-Water Inlet Pipe

Vent

Hot-Water Outlet Pipe

Draft Hood

Temperature-Pressure
Relief Valve

Baffle

Discharge Pipe

Flue

Anode Rod

Gas-Shutoff Valve

Dip Tube

Reset Button

Gas-Control Knob

Temperature-Control Dial

Drain Valve

Main Burner

Pilot

Chapter 7

Contents

How They Work

Gas-fired water heaters typically warm water to a temperature between 120°F and 140°F. When a hot-water faucet is opened, hot water flows from the top of the tank through the faucet, and cold water enters via the dip tube to replace it. Sensing a drop in temperature, a thermostat opens a valve that sends gas to the burner, where it is ignited by a pilot flame or an electric spark.

Combustion gases are vented from the burner chamber through the flue and its heat-retaining baffle, then out the draft hood and vent. In the unlikely event that the temperature or water pressure rises too high inside the water heater, a relief valve opens to prevent the tank from exploding. The anode rod is a magnesium bar that attracts impurities in the water which would otherwise attack the metal tank. By sacrificing itself to corrosion, it saves the rest of the tank from the same fate. A water heater may be joined to the gas line by a flexible connector or by a rigid pipe, as this one is.

Troubleshooting

Problem	Solution
• No hot water	Relight pilot light **122** •
• Pilot light does not stay lit	Tighten connections on thermocouple **124** • Replace a faulty thermocouple **124** •
• Not enough hot water	Stagger household use of hot water, raise the temperature, or use a larger water heater • Insulate tank and exposed hot-water line **127** • Clean the combustion chamber **127** •
• Water too hot	Lower the temperature, or have control unit inspected by a professional •
• Relief valve leaks	Test and replace the relief valve **129** •
• Drain valve leaks	Replace the drain valve **130** •
• Water heater tank leaks	Replace the water heater; no repairs are possible •
• Water heater makes rumbling noise	Drain sediment from the tank **128** •
• Hot water is dirty	Drain sediment from the tank **128** •
• Water in tank is rusty	Replace the anode rod **125** •

Before You Start

A problem with your water heater may be due to overwork, not mechanics.

A chronic shortage of hot water is often caused by heavy demand on a too-small heater. If your water heater holds less than about 15 gallons per family member (tank volume is stamped on a metal plate affixed to most water heaters), consider a larger unit or staggering your use of hot water.

Gas water heaters come in several designs. All have a flue-and-vent system that conducts outdoors dangerous carbon monoxide produced by burning gas. Clean the vent, baffle, and combustion chamber once a year to ensure good ventilation (*pages 126-127*).

Instead of a pilot flame to light the burner, some gas water heaters employ electronic ignition, which consumes less gas. If your water heater has electronic ignition, or if it has parts that differ from the model shown on page 118, seek the advice of a professional. Do the same for repairs to the control unit, pilot, burner, and gas-supply lines, or to replace a leaking or corroded tank.

Before You Start Tips:

⋯⋰ Your household will be without hot water while you work on the water heater. Inform your family in advance of any repair plans.

⋯⋰ Removal of parts such as drain valves and anode rods may require a helper to steady the water heater while you apply a wrench.

⋯⋰ Take old water-heater parts—as well as the brand and model number—with you when you buy replacements.

TOOLS

Open-end wrench
Vacuum cleaner
Wire brush
Soft brush
Old toothbrush
Pipe wrench
Socket wrench
Screwdriver
Water hose
Utility knife

MATERIALS

Thermocouple
Anode rod
Relief valve
Drain valve
Tank insulation
Duct tape

SAFETY FIRST

Before beginning any repairs, close the gas-shutoff valve. If you are uncomfortable working around gas appliances, call in a professional. Even if you do not plan to make your own repairs, know the location of the gas-shutoff valve in case of an emergency.

Lighting the Pilot

1. GAINING ACCESS TO THE PILOT

• Remove the burner access panels by lifting them off the heater *(right)* or sliding them sideways.

• To light a pilot that has blown out, turn the temperature-control dial to its lowest setting and the gas-control knob to OFF. Wait at least five minutes for the gas to clear. Then, if there is no gas odor, relight the pilot *(Step 2)*.

• If gas odor lingers, close the gas-shutoff valve supplying the water heater, ventilate the room, and call the gas company.

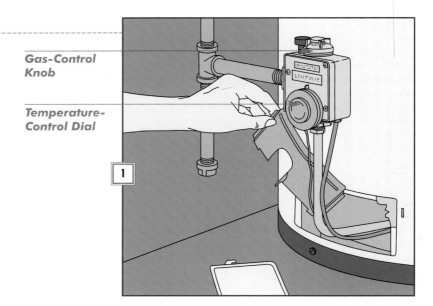

Gas-Control Knob

Temperature-Control Dial

1

2. SUPPLYING GAS TO THE PILOT

• Turn the gas-control knob to PILOT *(right).* Doing so supplies gas to the pilot burner when the reset button is pressed. As a safety precaution, gas is withheld from the main burner when the knob is in this position.

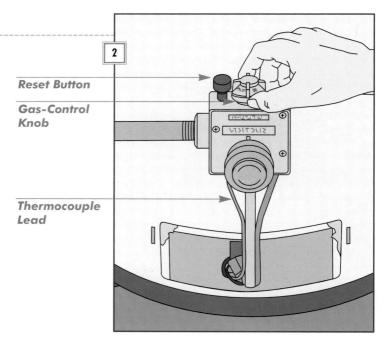

2

Reset Button

Gas-Control Knob

Thermocouple Lead

⚠ CAUTION

If you are heaving trouble lighting the pilot light, or if it won't stay lit, turn off the gas supply and call a plummer or the gas company for assistance.

Do not try to replace gas pipes or the control valve yourself. These are jobs for a professional.

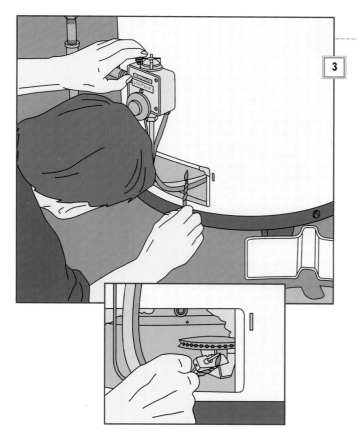

3

3. LIGHTING THE PILOT

● Hold down the reset button (or the gas-control knob if there is no reset button) while holding a lighted match near the pilot burner *(left)*. If the pilot is hard to reach, use long matchsticks or make a taper from a tightly rolled piece of paper.

● If the pilot fails to light after a few seconds, close the gas-shutoff valve and call the gas company. If it lights, continue to press the reset button or gas-control knob for one minute, then release it.

● If the pilot stays lit, proceed to Step 4. Otherwise, turn the gas-control knob to OFF. Try tightening the hexagonal nut connecting the thermocouple lead to the base of the control unit, first by hand, then by giving it a quarter-turn with an open-end wrench. Relight the pilot beginning with Step 2. If it goes out again, replace the thermocouple *(page 124)*.

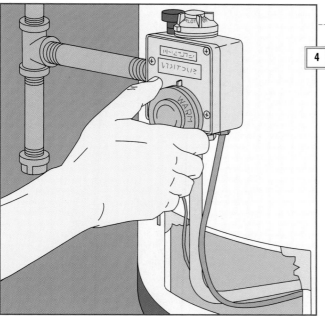

4

4. SETTING THE TEMPERATURE

● Turn the gas-control knob to ON; the main burner should light when the temperature-control dial is set between 120°F and 130°F, or just above WARM *(left)*. This moderate setting lowers heating costs, prolongs tank life, and reduces the risk of scalding.

● Replace the access panels.

Installing a Thermocouple

1. DISCONNECTING THE LEAD

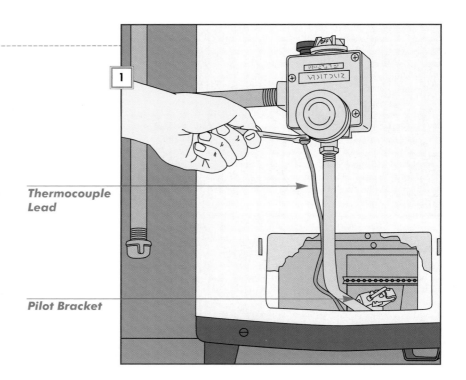

- Turn the gas-control knob to OFF and close the gas-shutoff valve. With an open-end wrench, loosen the nut that secures the thermocouple lead to the control unit *(right)*, then unscrew it by hand.

- Pull down on the copper tubing to detach the end of the thermocouple from the control unit. There may be a second nut attaching the thermocouple tip to the pilot bracket; unscrew the nut and slide it back along the copper lead.

- Grip the base of the thermocouple and slide it out of the pilot bracket.

Thermocouple Lead

Pilot Bracket

2. INSTALLING THE NEW THERMOCOUPLE

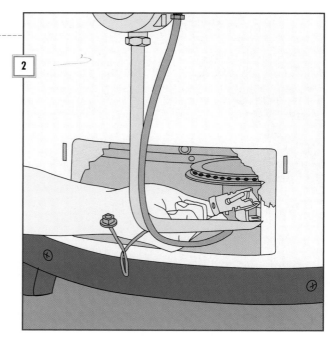

- Buy the proper replacement for the old thermocouple from a plumbing or heating supply store.

- Push the tip of the new thermocouple into the pilot bracket clip as far as it will go *(right)*. If there is a nut at its tip, screw it to the bracket.

- Uncoil the lead and gently form it into a curve. Screw the nut at the end of the lead to the control unit by hand, then give it a quarter-turn with an open-end wrench.

- Open the gas-shutoff valve, and relight the pilot *(page 122)*. If it goes out, close the gas-shutoff valve and call for service.

Replacing the Anode Rod

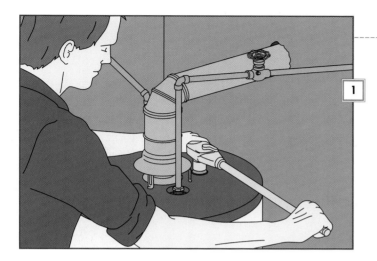

1

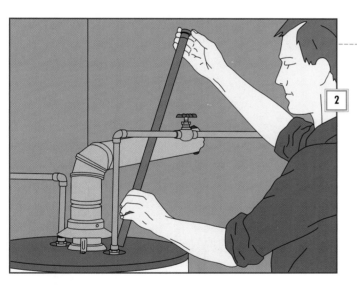

1. LOOSENING THE ANODE ROD

• Buy a replacement anode rod from a plumbing supply company.

• Close the cold-water-supply valve and turn the gas-control knob to OFF.

• Drain several gallons of water from the tank *(page 128)*. Loosen the anode rod with a pipe wrench or large socket wrench *(left)*, using steady pressure. Have a helper brace the tank to keep it from moving.

2. REPLACING THE ANODE ROD

• Lift the rod out of the tank *(left)*.

• Apply only a single width of pipe tape to the threaded upper end of the new rod. Insert the rod into the tank, screw it in as far as possible by hand, then tighten it with a wrench.

• Open the cold-water-supply valve and relight the pilot *(page 122)*.

2

Ron's TRADE SECRETS

REMOVING ANODE RODES

Lots of times, the anode rod I want to take out of a water heater is rusted tight to the top of the tank. A squirt of penetrating oil can help. Soak the joint two or three times, then give the oil several minutes to do its work. After this treatment, I find that a 3/4-inch-drive socket wrench will usually loosen up the rod, but if not, I reach for extra leverage. A steel pipe *(right),* slipped onto the wrench handle, generally does the trick.

Maintaining the Flue and Vent

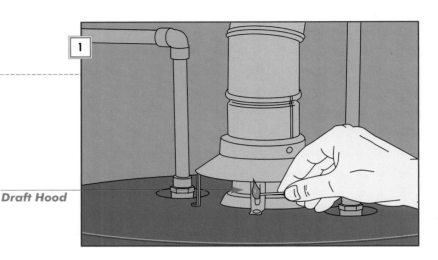

1. TESTING THE VENT

- Temporarily set the water heater at a higher temperature to light the burner.

- Wait 10 minutes, then hold a lighted match at the edge of the draft hood *(right)*. A properly working vent will draw the flame under the edge of the hood. If the flame is blown away from the hood or snuffed out, the vent may be blocked.

Draft Hood

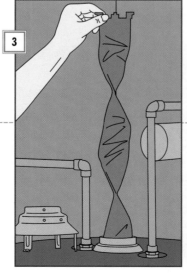

2. DISASSEMBLING THE VENT

- Turn the gas-control knob to OFF and close the gas-shutoff valve, then let the burner, draft hood, and vent cool.

- Remove the burner access panels *(page 122)*, and cover the burner and floor with newspapers to catch soot and debris.

- Mark the vent sections for reassembly. For support, wire any overhead ductwork to joists.

- Unscrew and remove the draft hood from the top of the tank *(right)*. Shake the hood and vent sections over the newspapers to release dirt, then scrub the insides gently with a wire brush.

- Replace rusted or perforated ductwork.

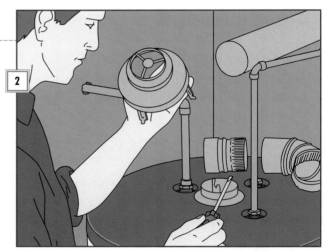

3. CLEANING THE BAFFLE

- With the vent removed, lift the baffle from the flue *(right)* and scrub it with a wire brush to remove dust and soot *(far right)*. If there is not enough room to pull the baffle all the way out, lift it as high as possible, cover the flue opening with a rag or newspaper, then brush the baffle and rattle it to dislodge debris.

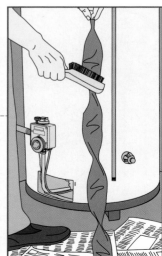

4. CLEANING THE COMBUSTION CHAMBER

● Reinstall the baffle, draft hood, and vent, then vacuum the inside of the combustion chamber *(left)*.

● Clean the burner and its ports with a soft brush. Use an old toothbrush to clean around the pilot, then vacuum the combustion chamber again and dispose of any newspapers.

● Relight the pilot *(page 122)*; test the vent with a lighted match as in Step 1. If the flame is not drawn up the vent, turn the gas-control knob to OFF and close the gas-shutoff valve. Recheck the pilot, burner and vent.

● Turn on the gas, relight the pilot, and test again. If the test fails, there may be a blockage in the main chimney—call for service.

Insulating a Water Heater

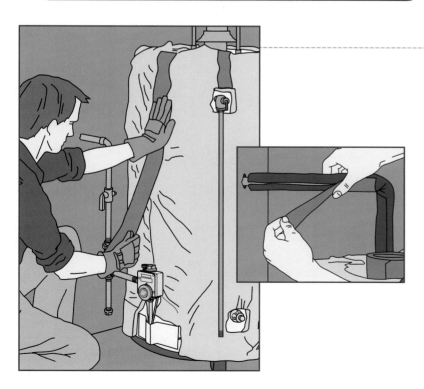

INSULATING THE TANK AND PIPES

Check whether the water-heater manufacturer recommends an insulation kit for your model.

● Turn off the gas-control knob and close the gas-shutoff valve. (On electric models, shut off power at the service panel.)

● Wrap insulation around the tank; fasten it to the tank top edge with duct tape. With a utility knife, cut insulation away from the access panels, relief and drain valves, control panel, and the space between the tank and floor *(left)*. Slit insulation to fit around the relief valve and discharge pipe.

● Tape any seams and start the heater *(page 122)*.

● Insulate runs of hot-water pipe with pre-slit foam tubes *(inset)*.

Draining the Tank

1. SHUTTING OFF THE WATER SUPPLY

• Turn the gas-control knob to OFF and close the gas-shutoff valve. (For an electric water heater, shut off power at the main service panel.)

• Close the cold-water-supply valve *(right)* and open a hot-water faucet somewhere in the house to speed draining.

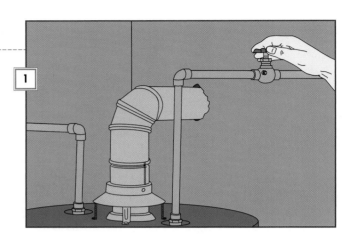

2. DRAINING AND FILLING THE TANK

• Attach a garden hose to the drain valve *(right)* and run it to a nearby floor drain, outdoors, or into a bucket. (Draining the tank by way of a bucket may take up to an hour.)

• Open the drain valve. As the tank empties, the valve may clog with sediment; open the cold-water-supply valve for a few minutes to allow water pressure to clear the blockage.

• To refill the tank, close the drain valve tightly, open the cold-water-supply valve and open any nearby hot-water faucet. When a steady stream of water flows from that faucet, the tank is full; close the faucet.

• Only after the tank is full, turn on the gas, then relight the pilot *(page 122)* or turn the power back on.

CONTROLLING SEDIMENT

Over time, sand, mineral scale, and other contaminants in the water supply build up as a layer of sediment on the bottom of a water tank, reducing the performance and shortening the life of both gas and electric water heaters. You can slow—but not stop—this process by softening the water and by lowering the temperature to 130°F or less. To keep sediment from accumulating, purge your tank every few months. To do so, drain off two or three gallons of water from the tank, then refill it as described above.

Installing a Relief Valve

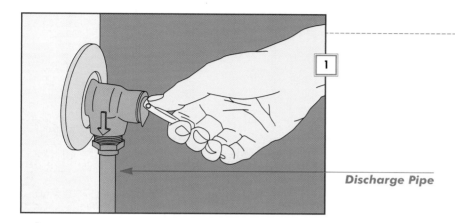

1. TESTING THE RELIEF VALVE

• Lift the spring lever on the valve *(left)* long enough for a cup or so of water to spurt out. Lift the lever several times to clear the valve of mineral scale. Keep clear of the discharge pipe as hot water escapes.

• If no water spurts out, or if water continues to drip after the valve is released, it should be replaced.

Discharge Pipe

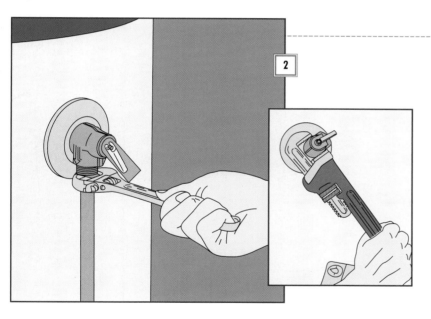

2. REMOVING THE RELIEF VALVE

• Turn the gas-control knob to OFF and close the gas-shutoff valve. (On an electric water heater, cut power at the main service panel.) Close the cold-water-supply valve.

• Drain a gallon of water from the tank if the relief valve is on top of the water heater, or five gallons if it is on the side *(page 128)*.

• Unscrew and remove the discharge pipe *(left)*, if there is one.

• Fit a pipe wrench over the relief valve *(inset)* and unscrew it. The valve may be difficult to remove. Use steady pressure and have a helper brace the tank. When the valve is loose, finish removing it by hand.

3. INSTALLING A NEW RELIEF VALVE

• Take the old valve with you to buy a replacement. Apply pipe tape to the threads of the new valve and screw it into the tank by hand, then tighten with a pipe wrench.

• Screw the discharge pipe (if any) into the valve outlet.

• Refill the water heater *(page 128)*, and relight the pilot *(page 123)* or turn on the electricity. If the valve leaks, have a plumber check for high water pressure in the house.

Removing a Drain Valve

A METAL VALVE

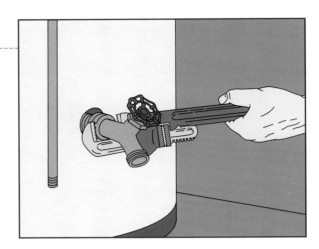

- Turn the gas-control knob to OFF and close the gas-shutoff valve. (For an electric water heater, shut off power at the main service panel.)

- Close the cold-water-supply valve and drain the heater completely *(page 128)*.

- Fit a pipe wrench over the base of the drain valve and unscrew the valve *(right)*.

A PLASTIC VALVE

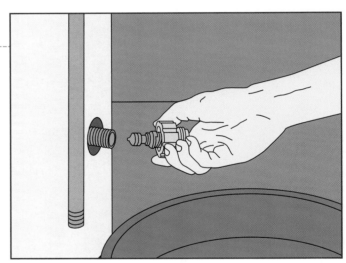

- Turn the valve handle counterclockwise by hand, four complete revolutions.

- While pulling firmly on the handle, turn it clockwise six complete turns to free it from the tank.

- Replace it with an identical valve. Insert the new valve, then push in on the handle and turn it counterclockwise six times, then clockwise four times. You can also replace a plastic valve with a sillcock valve *(page 131)*.

WHEN IS A DRAIN TRAY NECESSARY?

A drain tray is a watertight pan that fits under the water heater and is connected to drain piping. It is required by code when water heaters are located in attics and in some other locations. That's because a leaky fitting or even a small leak in the tank could cause significant damage over time unless the water heater is located on a concrete floor. The tray also catches water spurts from relief-valve tests. Even when not required, though, you may want to add a drain tray if the potential damage from a leak would be excessive, as in a basement shop. If you notice water in a drain tray, locate the source of the problem immediately.

Replacing a Drain Valve

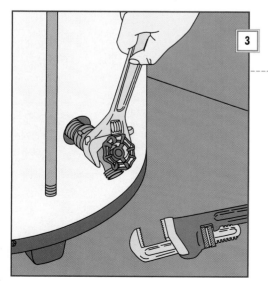

1. A NEW VALVE AND COUPLING

• Because it is both durable and contains a washer (and can therefore be repaired), a sillcock valve is the recommended replacement for both a metal valve and plastic one. Select a 3/4- or 1/2-inch valve with male threads; also buy a 3/4- to 1/2-inch reducing coupling to mate the 1/2-inch valve to the tank.

• Wrap pipe tape around the threaded end of the sillcock valve and screw it into the 1/2-inch end of the coupling *(left)*.

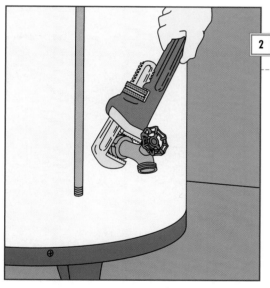

2. INSTALLING THE VALVE

• Apply pipe tape to the end of the nipple on the water heater tank.

• Screw the coupling and valve onto the nipple and tighten as far as possible by hand. Finish tightening the coupling with a pipe wrench.

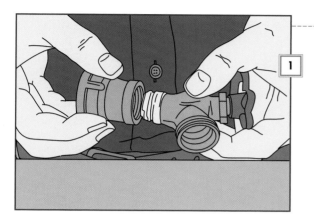

3. ALIGNING THE VALVE

• Fit an adjustable wrench over the body of the sillcock valve (but not over its outlet), and turn it clockwise to tighten the valve so that it faces down toward the floor.

• Refill the tank *(page 128)* and relight the pilot *(page 123),* or turn electricity back on.

FIX IT: Electric Water Heaters

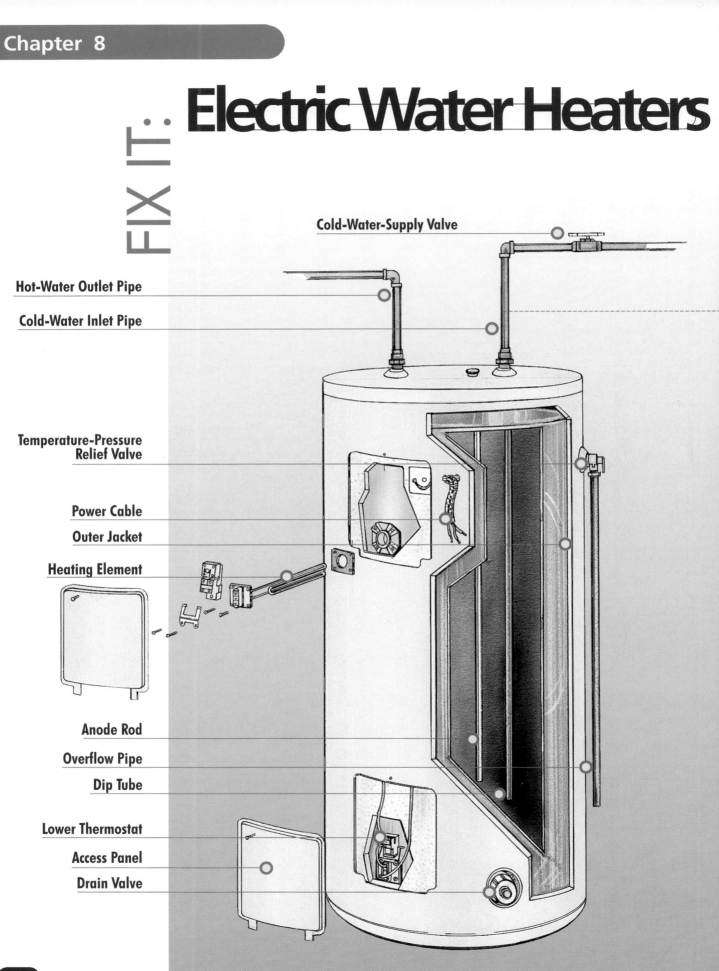

Cold-Water-Supply Valve

Hot-Water Outlet Pipe

Cold-Water Inlet Pipe

Temperature-Pressure Relief Valve

Power Cable

Outer Jacket

Heating Element

Anode Rod

Overflow Pipe

Dip Tube

Lower Thermostat

Access Panel

Drain Valve

Chapter 8

Contents

How They Work

An electric water heater shares many features with gas-heated units. In an electric model, however, curved rods called heating (or resistance) elements maintain water temperatures of between 120° and 140°F.

Electric water heaters usually have both an upper and a lower heating element, and each is controlled by a separate thermostat. When the thermostats sense a drop in the water's temperature, they close to complete an electrical circuit that includes the heating elements. The circuit is designed to prevent operation of both elements at the same time. When water in the tank reaches a suitable temperature, the thermostats open to break the circuit. The upper element has, in addition, a high-limit temperature cutoff to keep hot water at the top of the tank from reaching the boiling point.

A relief valve near or on top of the tank is a safety device. It relieves excessive pressures or temperatures by opening and releasing water through the overflow pipe.

Troubleshooting

Problem	Solution

No hot water
- Replace blown fuse or reset tripped breaker **136**
- Test and replace upper thermostat **137**
- Reset high-limit cutoff, or replace it if faulty **139**
- Test upper elements; clean or replace **140**

Not enough hot water
- Raise temperature setting on thermostats **137**
- Stagger household use of hot water
- Drain tank to remove sediment, then refill **128**
- Test and replace lower thermostat **137**
- Test and replace high-limit cutoff **139**
- Test and replace lower element **140**
- Clean scale from elements **141**
- Use a larger water heater *(see Before You Start)* **121**

Water too hot
- Lower temperature setting on thermostats **137**
- Test and replace thermostats **137**
- Test and replace high-limit cutoff **139**
- Test elements; clean or replace **140**

Circuit breakers (or fuses) trip repeatedly
- Call an electrician

Water heater leaks near elements
- Tighten loose element, or replace faulty gasket **140**

Water heater noisy
- Remove accumulated scale from heating elements **141**
- Drain sediment from the tank **128**

Relief valve leaks
- Test and replace the relief valve **129**

Drain valve leaks
- Replace the drain valve **130**

Water heater tank leaks
- Replace the water heater; no repairs are possible

Before You Start

An electric water heater is a simple and reliable appliance, whose basic maintenance and repair are the same as for gas water heaters (*Chapter 7*).

The Essential Difference

While testing the relief valve, replacing the anode, and draining the tank emphasize the similarities between the two types, everything that relates to heating the water is different because the source of heat is electricity.

To diagnose repairs of the heater's electrical system, you need a multitester. These inexpensive devices are available in home centers and electronics stores. They may be analog in design (with a dial) or digital. Both kinds work much alike.

For testing water heaters, you'll use the ohms scale most. A reading of 0 ohms indicates continuity (a completed circuit); a reading of infinity points to an open (incomplete) circuit. Always "center" the multitester before using it: Set the selector to RX1 and touch the probes together. On analog models, you'll also have to align the needle with 0.

Before You StartTips:

⋯⋗ Before disconnecting wires, label each with a masking tape "flag" as a guide to reconnecting them.

⋯⋗ To find the right water-heater temperature, turn the thermostat up or down 5° and live with the new temperature for a week before making additional adjustments.

TOOLS

Multitester
Screwdrivers
Socket wrench
Voltage tester

MATERIALS

Fuses
Masking tape
High-level cutoff
Thermostat
Heating element

SAFETY FIRST

Electric water heaters are 240-volt appliances whose exposed wires can deliver a fatal shock. Always turn off power to the water heater at the service panel before beginning work, verify that power has been shut off, and post a sign to warn others not to restore electricity prematurely. If you have any doubt whether electricity to a water heater has been turned off, seek professional assistance. Before restoring power, even to test a repair, replace any insulation that was moved and reattach access panels to prevent shock.

Checking for Power

CHECKING THE POWER SUPPLY

• If your water heater does not have its own disconnect box, check the main service panel for a blown fuse or tripped circuit breaker. Replace or reset them, if necessary. (A 240 volt heater may have two fuses or breakers.) If a water heater has its own disconnect box, turn it off and replace any blown fuses.

• If the heater still does not work, see if the circuit is receiving voltage *(right)*. The safest method is to hold the tip of an inductive voltage tester (power pen) near the wires in the disconnect box leading to the water heater. When the pen senses the presence of voltage, its indicator light glows (some also emit an audible signal).

• If no voltage is present even though the circuit is on, consult an electrician. When voltage is present, however, any problems are most likely with the water heater itself. Investigate the controls first *(page 137)*.

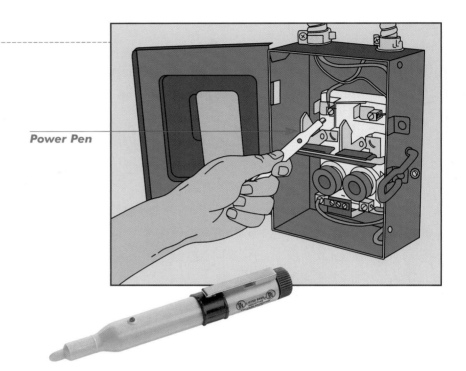

Power Pen

CHECKING CIRCUITS WITH A MULTITESTER

The power pen shown above is limited to checking for the presence or absence of power in a circuit. A multitester, however, is required for many of the tests in this chapter.

The tool checks electrical circuits for continuity, measures the amount of resistance (ohms) in the circuit, and can quantify the electrical force (volts) passing through a circuit. Unlike a power pen, it provides specific measurements, not generalized indications.

Because a multitester's probes must be in contact with wiring, the tool is best used on circuits after the power has been safely shut off. A digital multitester *(right)* is easy to use and is inexpensive.

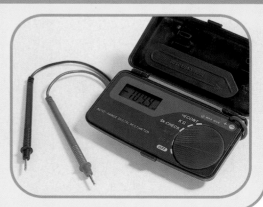

Getting at the Controls

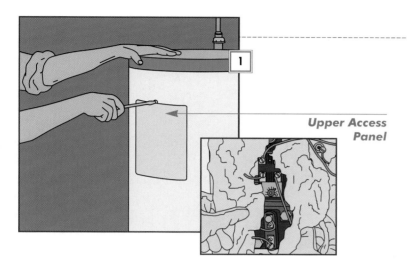

Upper Access Panel

1. REMOVING ACCESS PANELS

• Shut off power to the heater, and post a sign warning others not to turn it on.

• Unscrew and remove the upper access panel *(left)*; the lower access panel can be removed in the same way.

• Wearing gloves, carefully push insulation aside *(inset)* or lift it out without touching any wires or components. Save any insulation you remove.

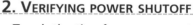

2. VERIFYING POWER SHUTOFF

• Touch the tip of a power pen to the upper terminals of the high-limit cutoff and to each incoming power supply wire *(left)*.

• If voltage is still present, check the circuit box again; you may not have tripped the correct breaker.

CAUTION: If you are unable to verify that power is off, do not work on the heater; call an electrician.

Troubleshooting Thermostats

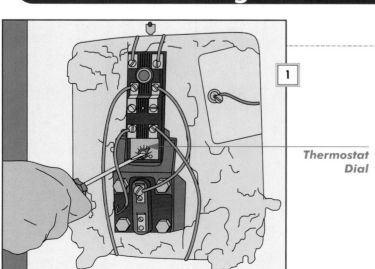

Thermostat Dial

1. ADJUSTING THE TEMPERATURE

• Turn off power to the water heater *(page 136)*; remove the upper access panel *(above)*.

• A water heater set too high wastes electrical energy, while one set too low won't provide enough hot water. Using a small screwdriver *(left)*, turn the thermostat dial counterclockwise to lower the temperature, or clockwise to raise it. If the water heater doesn't maintain the proper temperatures, test the thermostat *(Step 2)*.

2. TESTING THE UPPER THERMOSTAT

If either thermostat fails any of the following tests, replace it *(Step 4)*.

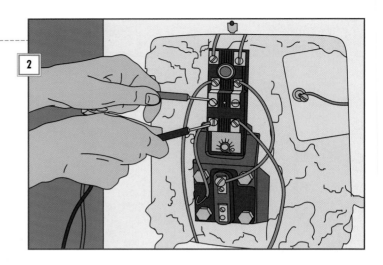

● Disconnect one wire to the upper element. Set a multitester to RX1 and touch a probe to each of the left thermostat terminals *(right)*. The tester should show infinity when the tank is warm.

● Touch a probe to each of the two right terminals (or to the upper left and upper right terminals on a 3-screw model). The tester should display 0.

● Adjust the thermostat to its highest setting. Repeat the two tests; the results should be reversed.

3. TESTING THE LOWER THERMOSTAT

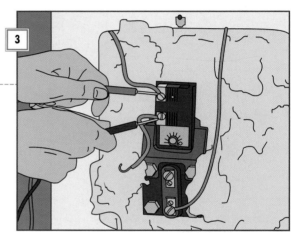

● Remove the lower access panel *(page 137)*.

● Turn the lower thermostat dial to its lowest setting and test as at right; the multitester (still at RX1) should read infinity when the tank is warm.

● Turn the dial to the highest setting; the needle should swing to 0.

● If any results differ, replace the faulty thermostat *(next step)*. If the thermostats test OK, test the elements *(page 140)*.

4. REMOVING THE THERMOSTAT

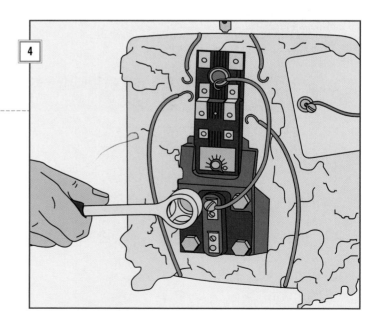

● Remove the cutoff *(page 139)*. Label and disconnect the wires to the thermostat.

● Using a socket wrench *(right)*, loosen the two bolts on the thermostat mounting bracket. Slip the thermostat up and out of the bracket.

● Buy a new thermostat of the same make and model at a heating or plumbing supplier. Before accepting the part, have the supplier check it for continuity.

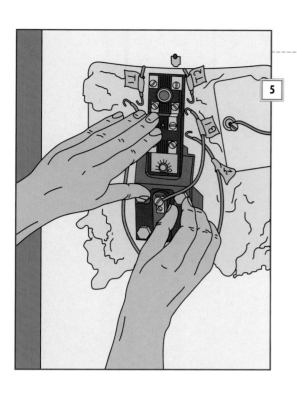

5. INSTALLING THE NEW THERMOSTAT

● Insert the new thermostat behind the bracket *(left)* and tighten the bolts.

● Adjust the thermostat or thermostats to the medium setting.

● Reinstall the cutoff *(page 139)*, reconnect all the wires, repack the insulation (making sure none gets behind the thermostat and cutoff), replace the access panels, and turn on the power. If the tank does not feel warm after three hours, test the heating elements *(page 140)*.

Testing the High-Limit Cutoff

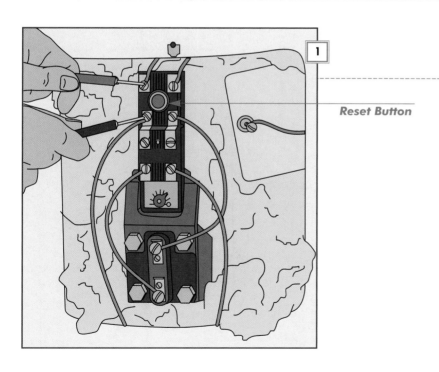

Reset Button

1. CHECKING THE RESET BUTTON AND TESTING THE CUTOFF FOR CONTINUITY

● Disconnect power to the heater *(page 136)*, then remove the upper access panel *(page 136)* and verify that power is off. Depress the reset button and listen for a click.

● If you hear no click, skip to the next paragraph. Otherwise, replace the insulation and access panel, then turn on the power. If you have warm water after 3 hours you have solved the problem. If the water is cold, turn off the power, and reopen the access panel.

● With a multitester set to RX1, touch the probes to the cutoff's left terminals *(left)*, and then to the right terminals. If the tester displays 0 each time, indicating continuity, test the thermostat *(page 138)*; if not, go to Step 2.

2. REMOVING THE CUTOFF

• Tag the wires to each of the cutoff terminals with masking tape to identify their positions for reassembly. Loosen the terminal screws and unhook the wires *(right)*.

• Remove the screws that hold the metal straps connecting the cutoff to the thermostat; the straps may be at the front or on the side. Take off the straps.

• Pull the cutoff upward to release it from spring clips that hold it to the heater, or gently pry it free with a screwdriver.

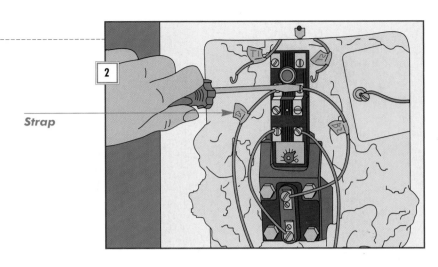

Strap

3. REPLACING THE CUTOFF

• Buy a new cutoff and have the supplier test it for continuity. Before installing it, depress the reset button.

• Snap the new cutoff into place and return wires, straps, insulation, and access panels to their original positions. Turn the power on. If the water is not warm after three hours, check the thermostats *(page 137)* and elements *(below)*.

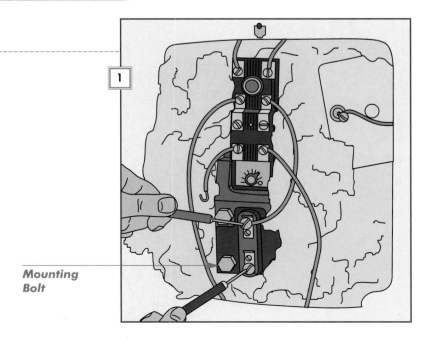

Replacing Heating Elements

1. CHECKING RESISTANCE

• Turn off power and remove the appropriate access panel *(page 136)*; upper and lower elements are tested in the same way.

• Disconnect one of the element wires.

• Set a multitester to RX1000. Touch one probe to an element mounting bolt and the other to each element terminal screw in turn. If the tester displays anything but infinity, replace the element *(Step 2)*.

• If the element passes the preceding test, set the multitester to RX1. Touch the probes to the terminal screws *(right)*. If there is any resistance reading at all, the element is good; otherwise, replace it.

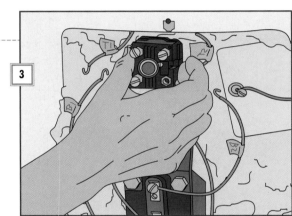

Mounting Bolt

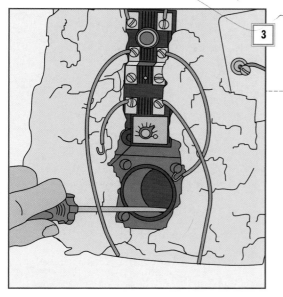

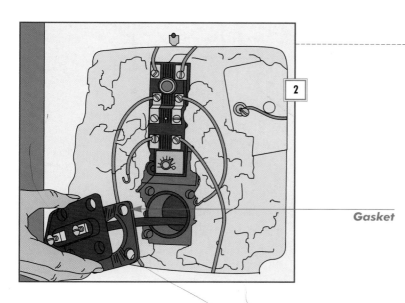

Gasket

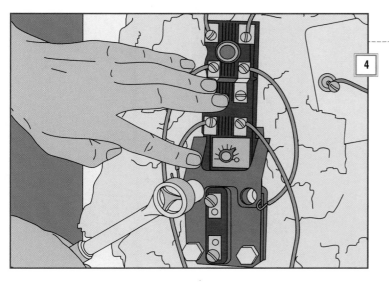

2. REMOVING THE ELEMENT

● Drain the heater *(page 128)*. Disconnect the remaining element wire.

● Remove the mounting bolts with a socket wrench (or use a socket wrench or pipe wrench to unscrew the element itself, depending on type).

● Pull the element straight out, gently working it loose *(left)*.

● If the element is faulty, purchase a replacement and a gasket to fit. If the element works but is being removed because the water heater is noisy, clean off mineral scale by soaking the element in vinegar for several hours, then chipping off the scale using an old knife.

3. CLEANING THE TANK FLANGE

● Using an old screwdriver, remove all remnants of the old gasket *(left)*. Scrape mineral scale and rust from the inside surface of the element fitting so the gasket can form a tight seal.

4. INSTALLING AN ELEMENT

● Place a new gasket onto the element (if the old element is being reused, remove all traces of the old gasket first).

● Ease the element into the heater, then tighten the element mounting bolts (or tighten the element itself, depending on type).

● Repack the insulation uniformly, reconnect the wires, replace the access panels, then turn the power back on.

Index

TIME® LIFE

Time-Life Books
is a division of Time Life Inc.

Time Life Inc.

George Artandi
President and CEO

Time-Life Books

Stephen R. Frary
President

Neil Kagan
Publisher/Managing Editor

How To Fix It:

Kitchen &
Bathroom Plumbing

Lee Hassig
Editor

Steve Schwartz
Marketing Director

Kate McConnell
Art Director / Series Designer

Wells P. Spence
Marketing Manager

Monika D. Lynde
Page Make-Up Specialist

Patricia Bray
Special Contributor (design)

Christopher Hearing
Director of Finance

Marjann Caldwell
Director of Book Production

Betsi McGrath
Director of Operations

John Conrad Weiser
Director of Photography
and Research

Barbara Levitt
Director of Editorial Administration

Marlene Zack
Production Manager

James King
Quality Assurance Manager

Louise D. Forstall
Library

Butterick Media

Staff for Kitchen &
Bathroom Plumbing

Mark D. Feirer
Editor

Caroline Politi
Director of Book Production

Jeff Beneke
Writer

David Joinnides
Page Layout

Jim Kingsepp
Technical Consultant

Annemarie McNamara
Copy Editor

Naomi Bibbins Bain
Editorial Coordinator

Lillian Esposito
Production Editor

Art Joinnides
President

Picture Credits

Fil Hunter
Cover Photograph

Bob Crimi
Geoff McCormack
Linda Richards
Joseph Taylor
Illustration

Brian Kraus
"Butterick Media"
Interior Photographs

First printing. Printed in U.S.A.
Published simultaneously in Canada.
School and library distribution
by Time-Life Education,
P.O. Box 85026, Richmond, Virginia 23285-5026.

TIME-LIFE is a trademark of Time Warner Inc. U.S.A.

**Library of Congress
Cataloging-in-Publication Data**
Kitchen & bathroom plumbing / by the editors of
Time-Life Books.
    p. cm.
Includes index.
    ISBN 0-7835-5650-0
    1. Kitchens—Maintenance and repair—Amaterus'
        manuals.
2. Bathrooms—Maintenance and repair—Amateurs'
        manuals.
3. Plumbing—Amateurs' manuals. I. Time-Life
Books.
TH6507.K59 1998   98-13019
696'.1—dc21   CIP